T0137097

Studies in Systems, Decision and Control

Volume 249

Series Editor

Janusz Kacprzyk, Systems Research Institute, Polish Academy of Sciences, Warsaw, Poland

The series "Studies in Systems, Decision and Control" (SSDC) covers both new developments and advances, as well as the state of the art, in the various areas of broadly perceived systems, decision making and control–quickly, up to date and with a high quality. The intent is to cover the theory, applications, and perspectives on the state of the art and future developments relevant to systems, decision making, control, complex processes and related areas, as embedded in the fields of engineering, computer science, physics, economics, social and life sciences, as well as the paradigms and methodologies behind them. The series contains monographs, textbooks, lecture notes and edited volumes in systems, decision making and control spanning the areas of Cyber-Physical Systems, Autonomous Systems, Sensor Networks, Control Systems, Energy Systems, Automotive Systems, Biological Systems, Vehicular Networking and Connected Vehicles, Aerospace Systems, Automation, Manufacturing, Smart Grids, Nonlinear Systems, Power Systems, Robotics, Social Systems, Economic Systems and other. Of particular value to both the contributors and the readership are the short publication timeframe and the world-wide distribution and exposure which enable both a wide and rapid dissemination of research output.

** Indexing: The books of this series are submitted to ISI, SCOPUS, DBLP, Ulrichs, MathSciNet, Current Mathematical Publications, Mathematical Reviews, Zentralblatt Math: MetaPress and Springerlink.

More information about this series at http://www.springer.com/series/13304

Jianxing Liu · Yabin Gao · Yunfei Yin ·
Jiahui Wang · Wensheng Luo ·
Guanghui Sun

Sliding Mode Control Methodology in the Applications of Industrial Power Systems

 Springer

Jianxing Liu
School of Astronautics
Harbin Institute of Technology
Harbin, China

Yabin Gao
School of Astronautics
Harbin Institute of Technology
Harbin, China

Yunfei Yin
School of Astronautics
Harbin Institute of Technology
Harbin, China

Jiahui Wang
College of Automation
Harbin Engineering University
Harbin, China

Wensheng Luo
School of Astronautics
Harbin Institute of Technology
Harbin, China

Guanghui Sun
School of Astronautics
Harbin Institute of Technology
Harbin, China

ISSN 2198-4182   ISSN 2198-4190 (electronic)
Studies in Systems, Decision and Control
ISBN 978-3-030-30657-1   ISBN 978-3-030-30655-7 (eBook)
https://doi.org/10.1007/978-3-030-30655-7

This Springer imprint is published by the registered company Springer Nature Switzerland AG
The registered company address is: Gewerbestrasse 11, 6330 Cham, Switzerland

To Yue, Jiahan and Jiaxuan
Jianxing Liu

To My Family
Yabin Gao

To My Family
Yunfei Yin

To My Family
Jiahui Wang

To My Family
Wensheng Luo

To My Family
Guanghui Sun

Preface

Sliding mode control (SMC) has been studied since the 1950s and widely used in practical applications, such as robot manipulators, aircraft, underwater vehicles, spacecraft, flexible space structures, electrical motors, automotive engines, power electronics systems, and fuel cell power systems, due to its insensitivity to parameter variations and robustness against external disturbances. Among these applications, many research endeavors have been focused during the past decades, on control and observation problems in fuel cell power systems and power electronics systems, using sliding mode technique. One of the most distinguished properties of SMC is that it utilizes a discontinuous control action, which switches between two distinctively different system structures such that a new type of system motion, called sliding mode, exists in a specified manifold. This particular characteristic of the motion in the manifold is its insensitivity to parameter variations, and complete rejection of external disturbances.

Polymer electrolyte membrane fuel cells (PEMFCs) have emerged as the most prominent technology for energizing future's automotive world. They are clean, quiet and efficient, and have been widely studied in automotive applications over the past two decades due to their relatively small size, lightweight and easy manufacture. While there are still major issues concerning cost, liability, and durability to be addressed before they become a widely used alternative to internal combustion engines (ICEs), fuel cells are expected to lead the world towards fossil-fuel-independent hydrogen economy in terms of energy and electro-mobility. One hindrance of fuel cells in general, as a kind of independent electrical power sources, is that their dynamic response is slow. Therefore, a fuel-cell-based power system requires additional storage elements with fast response time in order to handle rapid load variations.

This book is organized as follows. In Chap. 2, a brief review of sliding mode control for both linear and nonlinear systems are presented.

In Chap. 3, we focus on sliding mode observer (SMO) design for both linear uncertain systems and nonlinear systems. Two kinds of SMO design based on Utkin's and Lyapunov's methods for uncertain linear systems are recalled. Then, the SMO designs are extended to three forms of nonlinear systems (companion form, triangular input form and algebraical observable form). The applications of SMO based fault detection and isolation (FDI) have been presented and fault reconstruction via traditional SMOs is discussed. Then, our theoretical contribution in Second-Order SMO-based FDI is introduced. A novel adaptive SOSM Observer is developed and its application in FDI is presented. In the end, two illustrative examples are shown to both SMO designs and their applications in FDI.

In Chap. 4, the model of a PEMFC system which includes a stack voltage model and an air feed system model is discussed. The air feed system is modeled as a four-state model which considers the dynamics of oxygen partial pressure, nitrogen partial pressure, compressor speed, and supply manifold pressure. Then, a real-time PEMFC emulator is designed using experimental data obtained from a 33 kW PEMFC unit containing 90 cells in series. Finally, the proposed air feed system model is validated experimentally through the hardware-in-loop (HIL) test bench which consists of a physical air-feed system, based on a commercial twin screw compressor and a real-time PEMFC emulator.

In Chap. 5, we focus on the problem of robust control in PEMFC system, in order to maximize the fuel cell net power and avoid the oxygen starvation by regulating the oxygen excess ratio to its desired value during fast load variations. The oxygen excess ratio is estimated via an extended state observer from the measurements of the compressor flow rate, the load current and supply manifold pressure. An HIL test bench which consists of a commercial twin screw air compressor and a real-time fuel cell emulation system, is used to validate the performance of the proposed extended state observer (ESO)-based second-order sliding mode (SOSM) controller. The experimental results show that the proposed controller is robust and has a good transient performance in the presence of load variations and parametric uncertainties.

In Chap. 6, the estimation problem of the PEMFC system is presented. First, the design of an SOSM observer is presented for the PEMFC air-feed system, in order to estimate the hydrogen partial pressure in the anode channel of the PEMFC, using the measurements of stack voltage, stack current, anode pressure and anode inlet pressure. The robustness of this observer against parametric uncertainties and load variations is studied, and the finite time convergence property is proved via Lyapunov analysis. Then, an algebraical observer is designed for the partial pressures of oxygen and nitrogen in the cathode of the PEMFC. The states of the PEMFC air-feed system are presented in terms of a static diffeomorphism involving the system outputs (compressor flow rate and supply manifold pressure) and their time derivatives, respectively. The implementation of the algebraical observer on the HIL test bench is described. The effectiveness and robustness of the observer are validated experimentally.

In Chap. 7, we focus on the fault diagnosis problem of the PEMFC system, considering a fault scenario of sudden air leak in the air supply manifold. The design of an SOSM observer is presented for the PEMFC air-feed system, considering state estimation, parameter identification and fault reconstruction problems simultaneously. An adaptive-gain SOSM observer is developed for observing the system states, where the adaptive law estimates the uncertain parameters. The oxygen starvation phenomenon is monitored through an estimated performance variable (oxygen excess ratio). Satisfactory experimental results are obtained to show the effectiveness of the proposed observer. The effect of parameter variations and measurement noise are considered during the observer designs.

In Chap. 8, we concentrate on the control problem on the power side of the PEMFC system. Fuel cell power system requires power conditioning circuits with precise power control algorithms behind them such that the output power is compatible with the constraints of the power bus. These circuits include DC/DC converters which are used to control the voltages of the fuel cell and storage elements to the desired value. Furthermore, robust control design for DC/DC power converters is discussed, using sliding mode techniques. The main focus is on the necessary modification and improvement of conventional sliding mode methods for their applications to fuel cell power systems.

In Chap. 9, we consider the control and observation problems of three-phase AC/DC power converters. The controller design is based on the system model and has a cascaded structure which consists of two control loops. The outer loop regulates the DC-link capacitor voltage and the power factor providing the current references for the inner control loop. The current control loop tracks the actual currents to their desired values. To design the proposed controller, the load connected to the DC-link capacitor is considered as a disturbance, which directly affects the performance of the whole system. Theoretical analysis is given to show the closed-loop behavior of the proposed controller and experimental results are presented to validate the control algorithm under a real power converter prototype.

This book is a research monograph whose intended audience is graduate and postgraduate students, academics, scientists and engineers who are working in the field.

Harbin, China   Jianxing Liu
June 2019   Yabin Gao
                                                                                                      Yunfei Yin
                                                                                                   Jiahui Wang
                                                                                                 Wensheng Luo
                                                                                                 Guanghui Sun

Acknowledgements

There are numerous individuals without whose help this book will not have been completed. First, I would like to sincerely gratitude to Prof. Ligang Wu from Harbin Institute of Technology. It is really a great experience working with Prof. Wu's research group. With his abundant research experience, he shows me how to become an independent researcher. Our acknowledgments also go to our international collaborators, Prof. Leopoldo G. Franquelo, Dr. Sergio Vazquez, Dr. José I. Leon, Dr. Abraham Marquez, from Universidad de Sevilla, who have offered invaluable support and encouragement throughout this research effort. Thanks go to our students, Peng Chen, Chengwei Wu, Yongyang Xiong, Jiahui Wang, Tingting Yu, Huiyan Zhang, Xiaoning Shen, Tongyu Zhao, Hao Lin, Jinjiang Li, Tantan Dong.

Special thanks go to my supervisors Dr. Maxime Wack and Dr. Salah Laghrouche, for the continuous support of my Ph.D. study and research in Lab IRTES, for their patience, motivation, and immense knowledge, and for their inspiring guidance during my thesis. I would also like to thank all my colleagues in the laboratory for their help during my Ph.D. study. Especially, they are Fei Yan, Dongdong Zhao, Yishuai Lin, You Li, Jia Wu, You Zheng, Fayez Shakil Ahmed, Imad Matraji, Mohamed Harmouche, Adeel Mehmood.

Finally, I would like to express special thanks to my family. Without their endless love and understanding, I cannot finish this work.

Contents

Notations and Acronyms

$\triangleq$	Is defined as
$\sim$	Approximately equals to
$\ll$	Is much less than
$\gg$	Is much greater than
$\in$	Belongs to
$\forall$	For all
Σ	Sum
$\lvert \cdot \rvert$	Euclidean vector norm
$\lVert \cdot \rVert$	Euclidean matrix norm (spectral norm)
$\lVert \cdot \rVert_2$	$\mathcal{L}_2$-norm: $\sqrt{\int_0^\infty \lvert \cdot \rvert^2 dt}$ (continuous case)
	ℓ_2-norm: $\sqrt{\sum_0^\infty \lvert \cdot \rvert^2}$ (discrete case)
$\mathcal{L}_2\{[0,\infty),[0,\infty)\}$	Space of square summable sequences on $\{[0,\infty),[0,\infty)\}$ (continuous case)
$\ell_2\{[0,\infty),[0,\infty)\}$	Space of square summable sequences on $\{[0,\infty),[0,\infty)\}$ (discrete case)
$\dot{s}$	The first-order derivative of function s
$\ddot{s}$	The second-order derivative of function s
$\frac{\partial f}{\partial x}$ or $\frac{\partial}{\partial x}f$	The derivative of the function f with respect to x
$\mathbf{R}$	Field of real numbers
$\mathbf{R}^n$	Space of n-dimensional real vectors
$\mathbf{R}^{n \times m}$	Space of $n \times m$ real matrices
X^{T}	Transpose of matrix X
X^{-1}	Inverse of matrix X
$X > (<)0$	X is real symmetric positive (negative) definite
$X \geq (\leq)0$	X is real symmetric positive (negative) semi-definite
$*$	Symmetric terms in a symmetric matrix
0	Zero matrix
$0_{n \times m}$	Zero matrix of dimension $n \times m$
I	Identity matrix

I_n	$n \times n$ identity matrix
$\mathrm{col}\{x_1,\ldots, x_n\}$	Column vector $[x_1, \ldots, x_n]^{\mathrm{T}}$ with n elements
$\det(\cdot)$	The determinant computed from the elements of a square matrix
$\mathrm{diag}\{X_1,\ldots, X_{\mathrm{m}}\}$	Block diagonal matrix with blocks $X_1, \ldots, X_m$
inf	Infimum, the greatest lower bound
lim	Limit
$\ln(\cdot)$	The natural logarithm of a number
max	Maximum
min	Minimum
$\mathrm{rank}(\cdot)$	Rank of a matrix
$\mathrm{sing}(\cdot)$	The signum function of a real number
sup	Supremum, the least upper bound
$\lambda_{\min}(\cdot)$	Minimum eigenvalue of a real symmetric matrix
$\lambda_{\max}(\cdot)$	Maximum eigenvalue of a real symmetric matrix
AC	Alternating current
AFC	Aqueous alkaline fuel cell
AFE	Active front end
ARE	Algebraic Riccati equation
bar	A metric unit of pressure
DC	Direct current
DSP	Digital signal processor
Eq. or Eqs.	Equation or Equations
ESO	Eextended state observer
FC	Fuel cell
FDI	Fault detection and isolation
Fig. or Figs.	Figure or Figures
FNN	Fuzzy neural network
FOU	Footprint of uncertainty
FPGA	Field-programmable gate array
FTC	Fault tolerant control
GDM	Gradient descent method
HGO	High-gain observer
HIL	Hardware-in-loop
HOSM	High-order sliding mode
ICE	Internal combustion engine
IGBT	Insulated gate bipolar transistor
IT2	Interval type-2
IT2FNN	Interval type-2 fuzzy neural network
KF	Kalman filter
kVAr	Kilovolt-amperes reactive, thousand Volt-ampere reactive, a unit of reactive power
LMF	Lower membership function
LMI	Linear matrix inequality

LPV	Linear parameter varying
LTI	Linear time-invariant
MCFC	Molten carbonate fuel cell
PAFC	Phosphoric acid fuel cell
PEM	Proton exchange membrane
PEMFC	Proton exchange membrane fuel cell
PI	Proportional integral
PLL	Phase locked loop
PMSM	Permanent magnet synchronous
PR	Proportional plus resonant
PWM	Pulse-width modulation
THD	Total harmonic distortion
T-S	Takagi-Sugeno
SMC	Sliding mode control
SMO	Sliding mode observer
SOFC	Solid oxide fuel cell
SOSM	Second-order sliding mode
SOSML	The addition of linear term to the nonlinear SOSM term
SPD	Symmetrical positive definite
SRF	Synchronous reference frame
ST	Super-twisting
STA	Super-twisting algorithm
UIO	Unknown input observers
UKF	Unscented Kalman filter
VSC	Variable structure control

List of Figures

List of Tables

Chapter 1
General Introduction

This chapter introduces in detail the fuel-cell-based industrial power systems and several additional types of equipment required to make the fuel cell work at the optimal operating point. It also stresses some system failures or mechanical faults of the vulnerable power generation systems that can cause the shutdown or the permanent damage of the fuel cell. This monograph is focused on the polymer electrolyte membrane fuel cell air-feed system and power electronics systems. The motivation behind concentrating on these systems and the importance of their observation are discussed, and the outline of the monograph is described in this chapter.

1.1 Fuel-Cell-Based Industrial Power Systems

Polymer electrolyte membrane fuel cells (PEMFCs) have emerged as the most prominent technology for energizing future's automotive world. They are clean, quiet and efficient, and have been widely studied in automotive applications over the past two decades due to their relatively small size, light weight and easy manufacture [1, 5, 12, 13]. While there are still major issues concerning cost, liability and durability to be addressed before they become a widely used alternative to internal combustion engines (ICE), fuel cells are expected to lead the world towards fossil-fuel independent hydrogen economy in terms of energy and electro-mobility.

One hindrance of fuel cells in general, as a kind of independent electrical power sources, is that their dynamic response is slow. Therefore a fuel-cell-based power system requires additional storage elements with fast response time in order to handle rapid load variations. The most common elements are rechargeable batteries and super capacitors [15]. Hybrid electrical vehicles (HEV), such as Audi Q5, carry high-power batteries as well that can share the load with fuel cells. Obviously, their integration in the power system introduces additional converters in order to control their charging and discharging on the power bus. For example, let us consider a

© Springer Nature Switzerland AG 2020
J. Liu et al., *Sliding Mode Control Methodology in the Applications of Industrial Power Systems*, Studies in Systems, Decision and Control 249,
https://doi.org/10.1007/978-3-030-30655-7_1

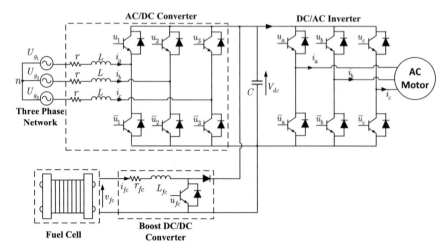

Fig. 1.1 PEMFC power system

typical fuel-cell based hybrid automotive power system, shown in Fig. 1.1. The power electronics topology has numerous interconnected components, i.e.

- Power elements (fuel cell stack and battery)
- A boost-type unidirectional DC/DC converter, for boosting fuel cell voltage to meet the power bus requirements
- A boost-type bidirectional DC/DC converter for connecting the battery to the bus
- A three-phase AC/DC rectifier, for occasional battery charging through external power source
- A three-phase DC/AC inverter for vehicle propulsion and traction motor(s)
- A multi-cell converter for DC loads (power windows, windshield wipers etc.).

It is clear that such a system requires sophisticated bus control algorithms in the background. In addition, automotive fuel cell applications have more rigorous operating requirements than stationary applications [3], as the risks of mechanical faults, leakage and explosion are higher in mobile systems. Technological malfunction can also turn the fuel-cell car into a potential hazard of life. These applications need precise control of performance, in order to guarantee reliability, health and safety of both, the fuel cell and the user. Along with control, health monitoring and safety systems are essential for the application of fuel cells in automobiles. These monitoring and control systems need to be precise, yet cost-effective and computationally simple, so that they can be easily implemented on commercial automotive embedded system platforms. Both control and health-monitoring systems require precise measurements of different physical quantities in the fuel cell. However, it is not always possible to use sensors for measurements, either due to prohibitive costs of the sensing technology or because the quantity is not directly measurable, especially in the conditions of humidified gas streams inside the fuel cell stack. In these cases, state observers serve as a replacement for physical sensors, for obtaining the unavailable

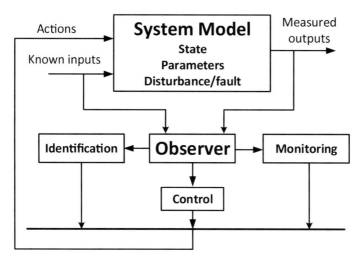

Fig. 1.2 Observer as the heart of control systems [2]

quantities, are of great interest. Observers form the heart of a general control problem and serve various purposes, such as identification, monitoring and control the system (Fig. 1.2). This book is dedicated to the problems associated with state observation in fuel cell systems.

1.2 Motivation and Features of the Book

We have seen that fuel-cell-based power generation systems are complex. Several additional types of equipment are required to make the fuel cell work at the optimal operating point. They are therefore vulnerable to system failures or mechanical faults that can cause the shutdown or the permanent damage of the fuel cell. Thus, reliable state observation and fault detection and isolation (FDI) schemes for such systems are necessary. As the entire fuel cell system cannot be covered in one study, the work carried out for this book is focused on the PEMFC air-feed system and power electronics systems. The motivation behind concentrating on these systems and the importance of their observation are discussed in the following subsections.

1.2.1 Sliding Mode Control, Observation and FDI

Sliding mode control (SMC) is a special kind of nonlinear control which has proven to be an effective robust control strategy for incompletely modeled or nonlinear systems since its first appearance in the 1950s [4, 8, 17]. One of the most distinguished properties of SMC is that it utilizes a discontinuous control action which switches between two distinctively different system structures such that a new type of system motion, called sliding mode, exists in a specified manifold. This particular characteristic of the motion in the manifold is its insensitivity to parameter variations and complete rejection of external disturbances. SMC has been applied primarily to the control of variable structure systems (VSS), its analysis and design are well presented in books, survey and tutorial papers [7, 8, 18, 19, 21], both from theoretical and implementation perspective.

SMC suffers from the so-called chattering phenomenon, which is undesirable because it often causes control inaccuracy, high heat loss in electric circuitry, and high wear of moving mechanical parts [21]. The chattering in SMC systems is usually caused by (1) the unmodeled dynamics with small time constants, which are often neglected in the ideal model; and (2) utilization of digital controllers with finite sampling rate, which causes so-called discretization chattering. Theoretically, the ideal sliding mode implies infinite switching frequency. Since the conventional SMC action is constant within a sampling interval, switching frequency cannot exceed that of half the sampling frequency, which leads to chattering as well. In practical applications, the high switching frequency is not possible or undesirable due to limitations of switching devices, such as switching losses, time delay, response time constant, presence of dead zone, hysteresis effect and saturation of device switching frequency [16]. In addition, the chattering action may excite the unmodeled high-order dynamics, which could damage actuators, systems and even leads to unforeseen instability [9]. Various approaches have been developed to address these problems, more details can be found in [6, 8, 18, 19, 21].

The idea of using a dynamical system to generate estimates of the system states was proposed by Luenberger in 1964 for linear systems [10]. FDI is usually achieved by generating residual signals, obtained from the difference between the actual system outputs and their estimated values calculated from dynamic models. The basic configuration of observer-based FDI is shown in Fig. 1.3.

Sliding mode techniques known for their insensitivity to parametric uncertainty and external disturbance, have been intensively studied and developed for observation and FDI problems, existing in the fuel cell power system. In particular, higher-order sliding mode (HOSM) approaches are considered as a successful technique due to the following advantages [11]:

- Possible to work with reduced order observation error dynamics
- Possible to estimate the system states in finite time
- Possible to generate continuous output injection signals
- Possible to offer chattering attenuation
- Robustness with respect to parametric uncertainties.

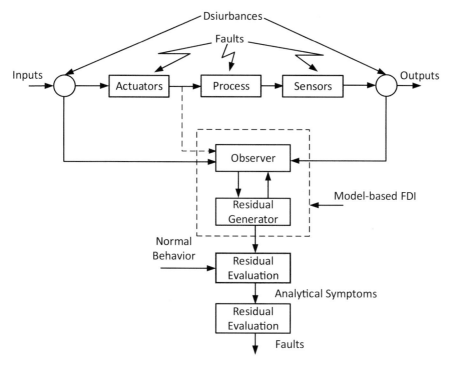

Fig. 1.3 Basic configuration of observer-based FDI [14]

1.2.2   PEMFC Power System

Fuel cells produce electricity through hydrogen and oxygen reaction. In PEMFCs, the anode and cathode sides are fed by hydrogen and oxygen, respectively. In fuel cell automobiles, hydrogen is stored in pressurized cylinders whereas air is used as an oxygen source, pumped into the cathode by a compressor. The air-feed system introduces an interesting challenge in the overall PEMFC system performance. As the PEMFC system works as an autonomous power plant in automobiles, the compressor motor is also powered by the PEMFC. Therefore, the net power of the system is the difference of the power produced by the fuel cell and that consumed by the air-feed system (consumption by the other auxiliary systems is negligible). Experimental studies have shown that the air-feed system can consume up to 30% of the fuel-cell power under high load conditions [20]. Therefore, it needs to be operated at its optimal point, at which it supplies just sufficient oxygen necessary for the hydrogen and oxygen reaction.

Unfortunately, such type of control is not possible without the knowledge of exact oxygen partial pressure. Conventional sensors can only give the total air pressure inside the cathode, which contains the partial pressures of other mixture gases such as nitrogen, carbon dioxide, etc. Imprecise knowledge of the oxygen quantity in the cathode can lead to serious problems, such as oxygen starvation during load transitions and hot-spots on the membrane surface [12]. To solve this problem, observers can be adopted. Observers can serve two important roles in the air-feed system. First, they can provide a precise estimate of the oxygen partial pressure. Second, they can detect any immediate variations in the nominal values of pressures throughout the air-feed system in order to identify anomalous behavior, thereby detecting and identifying faults and failures.

1.2.3 Power Electronics System

Fuel cell power systems require power conditioning circuits with precise power control algorithms behind them in order for the output power to be compatible with the constraints of the power bus. Power converters are key components in managing the energy flow through such hybrid systems. Control of power electronic systems requires the knowledge of several states, whereas practical systems are equipped with a limited number of voltage sensors due to cost concerns. An observer can augment the number of available-for-control states by using the output voltage measurements to observe the unknown current values. Along with the objective for control, they can also be used to identify system faults by detecting abnormal currents.

1.3 Outline of the Book

This book is organized as follows. In Chap. 2, a brief review of SMC for both linear and nonlinear systems are presented. The main contents of this monograph are shown in Fig. 1.4.

Chapter 3 firstly provides SMO designs for both linear uncertain systems and nonlinear systems. Two kinds of SMO design based on Utkin's and Lyapunov's methods for uncertain linear systems are recalled. Then, the SMO designs are extended to three forms of nonlinear systems (companion form, triangular input form and algebraical observable form). The applications of SMO based FDI are presented and fault reconstruction via traditional SMOs is discussed. Then, our theoretical contribution in Second Order SMO based FDI is introduced. A novel adaptive SOSM Observer is developed and its application in FDI is presented. In the end, two illustrative examples are shown concerning both SMO designs and their applications in FDI.

Chapter 4 describes the model of a PEMFC system which includes a stack voltage model and an air feed system model. The air feed system is modeled as a four-state model which considers the dynamics of oxygen partial pressure, nitrogen partial pressure, compressor speed, and supply manifold pressure. Then, a real-time PEMFC emulator is designed using experimental data obtained from a 33kW

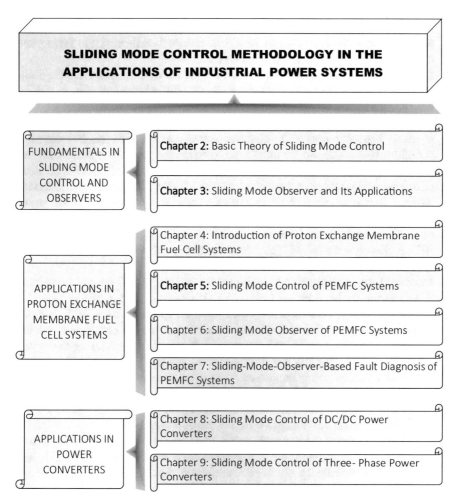

Fig. 1.4   Main contents of this publication

PEMFC unit containing 90 cells in series. Finally, the proposed air feed system model is validated experimentally through the HIL test bench which consists of a physical air-feed system, based on a commercial twin screw compressor and a real-time PEMFC emulator.

Chapter 5 focuses on the problem of robust control in PEMFC system, in order to maximize the fuel cell net power and avoid the oxygen starvation by regulating the oxygen excess ratio to its desired value during fast load variations. The oxygen excess ratio is estimated via an extended state observer from the measurements of the compressor flow rate, the load current and supply manifold pressure. An HIL test bench which consists of a commercial twin screw air compressor and a real-time fuel cell emulation system, is used to validate the performance of the proposed ESO-based SOSM controller. The experimental results show that the proposed controller is robust and has a good transient performance in the presence of load variations and parametric uncertainties.

Chapter 6 studied the estimation problem of the PEMFC system. First, the design of an SOSM observer is presented for the PEMFC air-feed system, in order to estimate the hydrogen partial pressure in the anode channel of the PEMFC, using the measurements of stack voltage, stack current, anode pressure, and anode inlet pressure. The robustness of this observer against parametric uncertainties and load variations is studied, and the finite time convergence property is proved via Lyapunov analysis. Then, an algebraical observer is designed for the partial pressures of oxygen and nitrogen in the cathode of the PEMFC. The states of the PEMFC air-feed system are presented in terms of a static diffeomorphism involving the system outputs (compressor flow rate and supply manifold pressure) and their time derivatives, respectively. The implementation of the algebraical observer on the HIL test bench is described. The effectiveness and robustness of the observer are validated experimentally.

Chapter 7 studied the fault diagnosis problem of the PEMFC system, considering a fault scenario of sudden air leak in the air supply manifold. The design of an SOSM observer is presented for the PEMFC air-feed system, considering state estimation, parameter identification and fault reconstruction problems simultaneously. An adaptive-gain SOSM observer is developed for observing the system states, where the adaptive law estimates the uncertain parameters. The oxygen starvation phenomenon is monitored through an estimated performance variable (oxygen excess ratio). Satisfactory experimental results are obtained to show the effectiveness of the proposed observer. The effect of parameter variations and measurement noise are considered during the observer designs.

Chapter 8 focuses on the power side of the PEMFC system. Fuel cell power system requires power conditioning circuits with precise power control algorithms behind them such that the output power is compatible with the constraints of the power bus. These circuits include DC/DC converters which are used to control the voltages of the fuel cell and storage elements to the desired value. Robust control design for DC/DC power converters is discussed, using sliding mode techniques. The main focus is on the necessary modification and improvement of conventional sliding mode methods for their applications to fuel cell power systems.

Chapter 9 considers the control and observation problems of three-phase AC/DC power converters. The controller design is based on the system model and has a cascaded structure which consists of two control loops. The outer loop regulates the dc-link capacitor voltage and the power factor providing the current references for the inner control loop. The current control loop tracks the actual currents to their desired values. To design the proposed controller, the load connected to the dc-link capacitor is considered as a disturbance, which directly affects the performance of the whole system. Theoretical analysis is given to show the closed-loop behavior of the proposed controller and experimental results are presented to validate the control algorithm under a real power converter prototype.

References

1. Al-Durra, A., Yurkovich, S., Guezennec, Y.: Study of nonlinear control schemes for an automotive traction PEM fuel cell system. Int. J. Hydrog. Energy **35**(20), 11291–11307 (2010)
2. Besançon, G.: Nonlinear Observers and Applications, vol. 363. Springer (2007)
3. Chen, F., Chu, H.S., Soong, C.Y., Yan, W.M.: Effective schemes to control the dynamic behavior of the water transport in the membrane of PEM fuel cell. J. Power Sources **140**(2), 243–249 (2005)
4. Emelyanov, S., Utkin, V.: Application of automatic control systems of variable structure for the control of objects whose parameters vary over a wide range. Dokl. Akad. Nauk SSSR **152**(2), 299–301 (1963)
5. Feroldi, D., Serra, M., Riera, J.: Design and analysis of fuel-cell hybrid systems oriented to automotive applications. IEEE Trans. Veh. Technol. **58**(9), 4720–4729 (2009)
6. Gao, W.: Fundamentals of Variable Structure Control Theory. Press of Science and Technology in China (in Chinese), Beijing (1990)
7. Hung, J.Y., Gao, W., Hung, J.C.: Variable structure control: a survey. IEEE Trans. Ind. Electron. **40**(1), 2–22 (1993)
8. Itkis, U.: Control Systems of Variable Structure. Wiley, New York (1976)
9. Kaynak, O., Erbatur, K., Ertugnrl, M.: The fusion of computationally intelligent methodologies and sliding-mode control–a survey. IEEE Trans. Ind. Electron. **48**(1), 4–17 (2001)
10. Luenberger, D.G.: Observing the state of a linear system. IEEE Trans. Mil. Electron. **8**(2), 74–80 (1964)
11. Perruquetti, W., Barbot, J.P.: Sliding Mode Control in Engineering. CRC Press (2002)
12. Pukrushpan, J.T., Stefanopoulou, A.G., Peng, H.: Control of Fuel Cell Power Systems: Principles, Modeling, Analysis and Feedback Design. Springer Science & Business Media (2004)
13. Raminosoa, T., Blunier, B., Fodorean, D., Miraoui, A.: Design and optimization of a switched reluctance motor driving a compressor for a PEM fuel-cell system for automotive applications. IEEE Trans. Ind. Electron. **57**(9), 2988–2997 (2010)
14. Saif, M., Xiong, Y.: Sliding mode observers and their application in fault diagnosis. In: Fault Diagnosis and Fault Tolerance for Mechatronic Systems: Recent Advances, pp. 1–57. Springer (2003)
15. Suh, K.W.: Modeling, analysis and control of fuel cell hybrid power systems. Ph.D. thesis, Department of Mechanical Engineering, The university of Michigan (2006)
16. Tan, S.C., Lai, Y.M., Tse, C.K.: Sliding Mode Control of Switching Power Converters: Techniques and Implementation. CRC Press (2011)
17. Utkin, V.: Variable structure systems with sliding modes. IEEE Trans. Autom. Control. **22**(2), 212–222 (1977)
18. Utkin, V., Gulder, J., Shi, J.: Sliding Mode Control in Electro-mechanical Systems. Automation and Control Engineering Series, vol. 34. Taylor & Francis Group (2009)
19. Utkin, V.I.: Sliding Modes in Control and Optimization. Springer, Berlin (1992)
20. Vahidi, A., Stefanopoulou, A., Peng, H.: Current management in a hybrid fuel cell power system: a model-predictive control approach. IEEE Trans. Control. Syst. Technol. **14**(6), 1047–1057 (2006)
21. Wu, L., Shi, P., Su, X.: Sliding Mode Control of Uncertain Parameter-Switching Hybrid Systems. Wiley (2014)

Chapter 2
Basic Theory of Sliding Mode Control

This chapter introduces some fundamentals of SMC including SMC design methods and main approaches to alleviate or limit the chattering. Moreover, some second-order SMC algorithms and definitions are introduced. It also presents the basics for the theoretical results developed in the subsequent chapters.

2.1 General Development of Sliding Mode Control

SMC is a special kind of nonlinear control which has proven to be an effective robust control strategy for incompletely modeled or nonlinear systems since its first appearance in the 1950s [15, 21, 35]. One of the most distinguished properties of SMC is that it utilizes a discontinuous control action which switches between two distinctively different system structures such that a new type of system motion, called sliding mode, exists in a specified manifold. This particular characteristic of the motion in the manifold provides insensitivity to parameter variations and complete rejection of external disturbances. The analysis and design of SMC have been well presented in the literature [19, 21, 36, 38], from both theoretical and implementation perspectives.

SMC, as one of the robust control strategies, is famous for its strong robustness to system uncertainties in a sliding motion. However, the uncertainties should satisfy the so-called 'matching' condition, that is, the uncertainties act implicitly within channels of the control input. If a system has mismatched uncertainties in the state matrix or/and the input matrix, the conventional SMC approaches are not directly applicable. Therefore, in the past two decades, many researchers investigated the SMC of uncertain systems with mismatched uncertainties/disturbances, see for example, [7, 8] and references therein. To mention a few, in [23], the SMC of uncertain second-order single input systems with mismatched uncertainties, was considered; In [8, 34], the authors investigated the SMC design for uncertain

© Springer Nature Switzerland AG 2020
J. Liu et al., *Sliding Mode Control Methodology in the Applications of Industrial Power Systems*, Studies in Systems, Decision and Control 249,
https://doi.org/10.1007/978-3-030-30655-7_2

systems, in which the uncertainties are mismatched and exist only in state matrix. The related approaches were then developed in [9, 10] to deal with a more complicated case that the mismatched uncertainties are involved in not only the state matrix but the input matrix. In addition, the integral SMC techniques were extensively used to deal with uncertain systems with mismatched uncertainties, see for example, [5, 6], and some other SMC approaches to deal with uncertain systems can be found in [13, 32].

In the past two decades, SMC has successfully been applied to a wide variety of practical systems such as robot manipulators, aircraft, underwater vehicles, spacecraft, flexible space structures, electrical motors, power systems, and automotive engines [11]. In general, SMC suffers from the so-called chattering phenomenon, which is undesirable because it often causes control inaccuracy, high heat loss in electric circuitry, and high wear of moving mechanical parts [38]. In addition, the chattering action may excite the unmodeled high-order dynamics, which could damage actuators, systems and even leads to unforeseen instability [22]. The chattering in SMC systems is usually caused mainly by the following reasons:

1. Utilization of digital controllers with finite sampling rate, which causes the so-called discretization chattering. Theoretically, the ideal sliding mode implies infinite switching frequency. Since the conventional SMC action is constant within a sampling interval, switching frequency cannot exceed that of half the sampling frequency, which leads to chattering.
2. The unmodeled dynamics with small time constants, which are often neglected in the ideal model.

In this section, we will present some preliminary background and fundamental theory of SMC at first, which is helpful to some readers who have no or limited knowledge on SMC, and then an overview of recent development of SMC methodologies and chattering reduction methods will be given.

2.2 Fundamental Theory of SMC

Let us consider the following nonlinear system:

$$\dot{x}(t) = f(x, t) + g(x, t)u(t), \tag{2.1}$$

where $x \in \mathbf{R}^n$ is the state variable vector, $u(t) \in \mathbf{R}^m$ is the control input, $f(\cdot, \cdot)$ and $g(\cdot, \cdot)$ are continuous functions in x, u and t vector fields [12, 36].

The sliding mode controller

$$u(t) = \begin{bmatrix} u_1(t) \ u_2(t) \ \cdots \ u_m(t) \end{bmatrix}^{\mathrm{T}} \tag{2.2}$$

is designed as

$$u_i(t) = \begin{cases} u_i^+(t), & \text{if } s_i(x) > 0, \\ u_i^-(t), & \text{if } s_i(x) < 0, \end{cases} \quad i = 1, 2, \ldots, m, \tag{2.3}$$

where $u_i^+(t) \neq u_i^-(t)$, and $s(x) \in \mathbf{R}^m$ called sliding manifold is the switching vector function $s(x) = \begin{bmatrix} s_1(x) & s_2(x) & \cdots & s_m(x) \end{bmatrix}^\mathrm{T}$. It undergoes discontinuities on the surface $s(x) = 0$.

Note that the SMC law (2.3) is designed to ensure that the sliding surface $(s(x) = 0)$ is reached and then motion on the sliding surface is maintained. This means that the so-called 'reachability condition' should be satisfied by manipulating the control law $u(t)$. The necessary and sufficient condition for the system (2.1) to satisfy the reachability condition is expressed as

$$s(x)\dot{s}(x) < 0. \tag{2.4}$$

Condition (2.4) guarantees that the trajectory of the system states always points towards the sliding surface. A more strict reachability condition called 'η-condition' is given as follows

$$s(x)\dot{s}(x) \leq -\eta|s(x, t)|, \tag{2.5}$$

where η is a positive scalar. Condition (2.5) ensures that the sliding surface is reached in finite time despite the presence of uncertainty. The system state trajectories under the reaching condition (2.4) or (2.5) are called the reaching phases [12, 19].

2.2.1 SMC Design Methods

Several SMC design methods have been proposed in literature which mainly consist of two steps [12, 36]:

Step 1: Design a sliding manifold $s(x)$ which provides desired performance in the sliding mode, such as stability, disturbance rejection capability and tracking;

Step 2: Design a discontinuous feedback control $u(t)$ which will force the system states to reach the sliding manifold in finite time, thus the desired performance is attained and maintained.

For ease of implementation, the sliding variable $s_i(x)$, $i = 1, 2, \ldots, m$ is chosen as a linear combination of the state variables, expressed as

$$s_i(x) = \sum_{j=1}^{m} \alpha_j x_j(t), \tag{2.6}$$

where α_j denotes the sliding coefficients and $x_j(t) \in x(t)$. The main objective of the sliding mode controller is to drive the system state trajectories onto the specified sliding surface in a finite time and maintained there for all subsequent time. Typical SMC strategies will be introduced in the following part.

Equivalent-Control-Based Design

For the system (2.1), assuming that the term $\frac{\partial s}{\partial x} g(x, t)$ is non-singular, the control law $u(t)$ is designed as follows,

$$u(t) = u_{eq}(t) + u_N(t), \tag{2.7}$$

where $u_{eq}(t)$ represents a continuous component and $u_N(t)$ represents a discontinuous component.

The equivalent control $u_{eq}(t)$ is derived from the so-called *equivalent control method*, i.e., in the case when $s(x) = \dot{s}(x) = 0$. Thus, $u_{eq}(t)$ is calculated as

$$u_{eq}(t) = -\left(\frac{\partial s}{\partial x} g(x, t)\right)^{-1} \frac{\partial s}{\partial x} f(x, t). \tag{2.8}$$

Substituting the above equivalent control (2.8) into the original system (2.1), it follows that the motion of sliding mode is determined by

$$\dot{x}(t) = \left[I_n - g\left(\frac{\partial s}{\partial x} g(x, t)\right)^{-1} \frac{\partial s}{\partial x} \right] f(x, t), \tag{2.9}$$

where (2.9) is considered as the equation of the sliding mode in the manifold $s(x) = 0$. The high frequency switching action $u_N(t)$ can be designed as

$$u_N(t) = -\beta \left(\frac{\partial s}{\partial x} g(x, t)\right)^{-1} \text{sign}(s(x)), \quad \beta > 0 \tag{2.10}$$

such that the derivative of the Lyapunov function $V = \frac{1}{2} s^T(x) s(x)$ is negative, that is

$$\dot{V} = s^T(x) \dot{s}(x) = s^T(x) \frac{\partial s(x)}{\partial x} g(x, t) u_N(t) < -\beta \|s(x)\|. \tag{2.11}$$

Remark 2.1 The physical meaning of the equivalent control can be interpreted as the low-frequency component of the discontinuous control law $u(t)$, because the high-frequency $u_N(t)$ can be filtered out by a low pass filter of the system

$$\tau \dot{z} + z = u(t), \quad \tau \ll 1, \tag{2.12}$$

which means $z \simeq u_{eq}$.

Reaching Law Approach

The reaching law specifies the dynamics of a switching function, which can be described by the following differential equation:

$$\dot{s}(x) = -\varUpsilon \operatorname{sign}(s(x)) - Kg(s(x)), \qquad (2.13)$$

where

$$\varUpsilon = \operatorname{diag}\{\varepsilon_1, \varepsilon_2, \dots, \varepsilon_m\}, \quad \varepsilon_i > 0,$$
$$K = \operatorname{diag}\{k_1, k_2, \dots, k_m\}, \quad k_i > 0,$$

$$\operatorname{sign}(s(x)) = \begin{bmatrix} \operatorname{sign}(s_1(x)) \\ \operatorname{sign}(s_2(x)) \\ \vdots \\ \operatorname{sign}(s_m(x)) \end{bmatrix}, \quad g(s(x)) = \begin{bmatrix} g_1(s_1(x)) \\ g_2(s_2(x)) \\ \vdots \\ g_m(s_m(x)) \end{bmatrix},$$

$$g_i(0) = 0, \quad s_i(x)g_i(s_i(x)) > 0, \quad i = 1, \dots, m.$$

Equation (2.13) is a general form of reaching law, and some special cases are

1. The constant rate reaching law:

$$\dot{s}(x) = -\varUpsilon \operatorname{sign}(s(x)).$$

2. The constant plus proportional rate reaching law:

$$\dot{s}(x) = -\varUpsilon \operatorname{sign}(s(x)) - Ks(t).$$

3. The power rate reaching law:

$$\dot{s}_i(x) = -\varepsilon_i |s_i(x)|^\alpha \operatorname{sign}(s_i(x)), \quad 0 < \alpha < 1.$$

The reaching law approach not only guarantees the reaching condition but also specifies the dynamics of the motion during the reaching phase [19].

2.2.2 Chattering Problem

Chattering problem is one of the main obstacles for applying SMC to real applications. The chattering in SMC systems is usually caused by (1) the unmodeled dynamics with small time constants, which are often neglected in the ideal model; and (2) utilization of digital controllers with finite sampling rate, which causes so called 'discretization chattering'. From control engineers' point of view, chattering is undesirable because it often causes control inaccuracy, high heat loss in electric circuitry, and high wear of moving mechanical parts. In addition, the chattering action

Table 2.1 Main approaches to alleviate or limit the chattering problems

Approach	Operation principle
Boundary layer approach	To insert a boundary layer near the sliding surface so that a continuous control action replaces the discontinuous one when the system is inside the boundary layer [33]. For this purpose, the discontinuous component of the controller: $u_N(t) = -K_s \operatorname{sign}(s(x))$ is often replaced by the saturation control: $u_N(t) \approx -K_s \dfrac{s(x)}{\|s(x)\| + \delta}$ for some, preferably small, $\delta > 0$. The boundary layer approach has been utilized extensively to the practical applications. However, this method has some disadvantages such as: • It may give a chattering-free system but a finite steady-state error must exist • The boundary layer thickness has the trade-off relation between control performance of SMC and chattering mitigation • Within the boundary layer, the characteristics of robustness and the accuracy of the system are no longer assured
Reaching law approach	Since the amplitude of chattering depends on the magnitude of control, the intuitive way of chattering reduction is to decrease the amplitude of the discontinuous control [17]. This technique affects the robustness property of the controller and degrades the transient response of the system. Therefore, exists a trade-off between the chattering reduction and the system performance. A compromised approach is to decrease the amplitude of the discontinuous control, $u_N(t)$, when the system state trajectories are near to sliding surface (to reduce the chattering), and to increase the amplitude when the system states are far from the sliding surface
Dynamic SMC approach	The main idea of the dynamic SMC approach is to insert an integrator (or any other strictly proper low-pass filter) between the SMC and the controlled plant [4]. The concept is illustrated here: The time derivative of the control input, ω, is treated as the new control input for the augmented system. Since the low-pass integrator filters out the high frequency chattering in ω, the control input to the real plant, u, becomes continuous which offers a possibility to reduce chattering. Such a method can eliminate chattering and ensure zero steady-state error, however, it should be noted that the system order is increased by one and the transient responses must be degraded [37]

may excite the unmodeled high-order dynamics, which probably leads to unforeseen instability [24]. Therefore, various methods have been proposed in literature to reduce or soften the chattering action [1, 4]. Among others, the main approaches to avoid or limit the chattering problems are shown in the Table 2.1.

2.3 SOSM Control

Apart from the above-mentioned chattering elimination/reduction approaches, HOSM technique is an effective method for chattering attenuation [3, 14, 25]. In this technique, the discontinuous control is applied on a higher time derivative of the sliding variable, such that both the sliding variable and its higher time derivatives converge to the origin. It is of great interest to see that the discontinuous control does not act upon the system input directly, and the real control input single is continuous, therefore, chattering effect is alleviated while the properties of robustness and finite time convergence of classical SMC are retained. Among various HOSM control techniques, SOSM control technique is the most popular and effective method which forces the sliding variable and its first order time derivative to zero in finite time [3, 26–28].

2.3.1 Definitions

Consider a discontinuous differential equation in the sense of Filippov [16]

$$\dot{x} = v(x), \tag{2.14}$$

where $x \in X \subset \mathbb{R}^n$ is the state vector, v is a locally bounded measurable (Lebesgue) vector function. The equation can be replaced by an equivalent differential inclusion

$$\dot{x} \in \mathbb{V}(x). \tag{2.15}$$

If the vector-field v is continuous almost everywhere, $\mathbb{V}(x)$ is the convex closure of the set of all possible limits of $\mathbb{V}(y)$ as $y \to x$, while $\{y\}$ are continuity points of v. Solutions of the equation are defined as absolutely continuous functions $x(t)$, satisfying the differential inclusion almost everywhere [31]. Let a constraint function given by

$$s(t, x(t)) = 0, \tag{2.16}$$

where $s : \mathbb{R}^n \to \mathbb{R}$ is a sufficiently smooth function.

Definition 2.1 ([31]) Suppose that

- Successive total time derivatives s and $\dot{s}$ are continuous functions of the system state variables. In other words, the discontinuity does not appear in the $\dot{s}$;
- The set $s = \dot{s} = 0$ is non-empty and consists of Filippov's trajectories [31].

Then, the motion on the set $s = \dot{s} = 0$ is called a second sliding mode with respect to the constraint function s (Fig. 2.1).

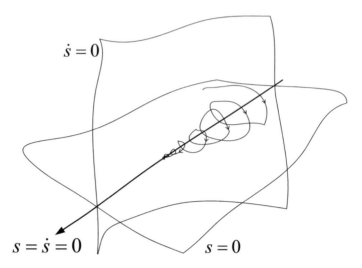

Fig. 2.1   SOSM trajectory

2.3.2   SOSM Dynamics

Consider the following nonlinear system [25]

$$\dot{x}(t) = f(t, x(t), u),$$

$$s(t) = s(t, x),$$

(2.17)

where $x \in X \subset \mathbb{R}^n$ is the state vector, $u \in U \subset \mathbb{R}$ is the bounded control input, t is the independent variable time and f is a sufficiently smooth uncertain vector function. The control objective is to force the sliding variable and its time derivative $s(t)$ and $\dot{s}(t)$ to zero in finite time, i.e.

$$s(t) = \dot{s}(t) = 0.$$

(2.18)

Assume that the control task is fulfilled by its zero dynamics with respect to the sliding variable $s(t, x)$. By differentiating the sliding variable s twice,

$$\dot{s}(t) = \frac{\partial}{\partial t} s(t, x) + \frac{\partial}{\partial x} s(t, x) f(t, x, u),$$

$$\ddot{s}(t) = \underbrace{\frac{\partial}{\partial t}\dot{s}(t, x, u) + \frac{\partial}{\partial x}\dot{s}(t, x, u) f(t, x, u)}_{\varphi(t,x)} + \underbrace{\frac{\partial}{\partial u}\dot{s}(t, x, u)\,\dot{u}(t)}_{\gamma(t,x)}.$$

(2.19)

Depending on the relative degree [20] of the nonlinear SISO system (2.17), two cases are considered

- **Case a**: relative degree $r = 1$, i.e., $\frac{\partial}{\partial u}\dot{s} \neq 0$;
- **Case b**: relative degree $r = 2$, i.e., $\frac{\partial}{\partial u}\dot{s} = 0$, $\frac{\partial}{\partial u}\ddot{s} \neq 0$.

Relative Degree 1

In this case, the control problem can be solved by the classical first order SMC, nevertheless second order SMC can be used in order to avoid chattering. Shortly speaking, the time derivative of the control $\dot{u}(t)$ may be considered as the actual control variable. A discontinuous control $\dot{u}(t)$ steers the sliding variable s and its time derivative $\dot{s}$ to zero, so that the plant control u is continuous and the chattering is avoided [4, 25].

The second time derivative $\ddot{s}$ (2.19) is described by the following equation

$$\ddot{s} = \varphi(t, x) + \gamma(t, x)\dot{u}(t), \tag{2.20}$$

where $\varphi(t, x)$ and $\gamma(t, x)$ are some bounded functions. The following conditions are assumed [25]:

1. The control values belong to the set $U = \{u : |u| \leq U_M\}$, where $U_M = $ Constant > 1.
2. There exists $u_1 \in (0, 1)$ such that for any continuous function $u(t)$ with $|u(t)| > u_1$, there is t_1, such that $s(t)u(t) > 0$ for each $t > t_1$.
3. There exist positive constants s_0, K_m and K_M such that if $|s(t, x)| < s_0$, then

$$0 < K_m < \gamma(t, x) < K_M, \ \forall u \in U, \ x \in X, \tag{2.21}$$

and the inequality $|u| > u_0$ entails $\dot{s}u > 0$.
4. There exists constant C such that within the region $|s(t, x)| < s_0$ the following inequality holds,

$$|\varphi(t, x)| \leq C, \ \forall u \in U, \ x \in X. \tag{2.22}$$

Condition 2 means that there exists a proper control $u(t)$ forcing the sliding variable into a set for any initial value of state, given that the boundedness conditions on the sliding dynamics defined by Conditions 3 and 4 are satisfied. It follows from (2.20), (2.21) and (2.22) that all solutions satisfy the differential inclusion

$$\ddot{s} \in [-C, C] + [K_m, K_M]\dot{u}(t). \tag{2.23}$$

Relative Degree 2

In this case, the control problem statement can be derived by considering the variable u as a state variable and $\dot{u}$ as the actual control. Suppose the system dynamics (2.17) is affine in the control law, i.e.,

$$f(t, x(t), u) = a(t, x) + b(t, x)u, \tag{2.24}$$

where $a : \mathbb{R}^{n+1} \to \mathbb{R}^n$ and $b : \mathbb{R}^{n+1} \to \mathbb{R}^n$ are sufficiently smooth uncertain vector functions. Equation (2.19) can be rewritten as

$$\dot{s}(t) = \frac{\partial}{\partial t} s(t, x) + \frac{\partial}{\partial x} s(t, x) a(t, x) + \frac{\partial}{\partial x} s(t, x) b(t, x) u \tag{2.25}$$

$$= \frac{\partial}{\partial t} s(t, x) + \frac{\partial}{\partial x} s(t, x) a(t, x),$$

$$\ddot{s}(t) = \frac{\partial^2}{\partial t^2} s(t, x) + \frac{\partial}{\partial x} s(t, x) \frac{\partial}{\partial t} a(t, x)$$

$$+ \left[\frac{\partial^2}{\partial t \partial x} s(t, x) + a^{\mathrm{T}}(t, x) \frac{\partial^2}{\partial x^2} s(t, x) + \frac{\partial}{\partial x} s(t, x) \frac{\partial}{\partial x} a(t, x) \right]$$

$$\times \left[a(t, x) + b(t, x) u(t) \right]$$

$$= \varphi(t, x) + \gamma(t, x) u(t).$$

It follows from (2.21), (2.22) and (2.25) that all solutions satisfy the following differential inclusion

$$\ddot{s} \in [-C, C] + [K_m, K_M] u(t). \tag{2.26}$$

2.3.3 SOSM Controllers

In this part, the most well known SOSM controllers are introduced, i.e., the super-twisting controller, the twisting controller and the sub-optimal controller. These controllers are insensitive to some model perturbations and external disturbances. Given that the expression for the sliding manifold is known, it is possible to design the constant parameters of the controllers [31].

Super-Twisting Control Algorithm

The super-twisting algorithm (STA) is a unique absolutely continuous sliding mode algorithm ensuring all the main properties of first order sliding mode control for system with Lipschitz continuous matched uncertainties/disturbances with bounded gradients [18]. The STA was developed to control systems with relative degree one in order to avoid chattering in variable structure control (VSC). The trajectories on the second sliding manifold are shown in Fig. 2.2.

Consider the system (2.23), the control algorithm is defined as follows [25]

Fig. 2.2 STA phase
trajectory

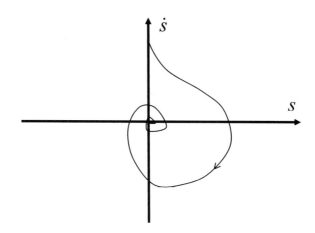

$$u(t) = u_1(t) + u_2(t),$$

$$\dot{u}_1(t) = \begin{cases} -u, & \text{if } |u| > 1, \\ -\alpha \text{sign}(s), & \text{otherwise}, \end{cases}$$

$$u_2(t) = \begin{cases} -\lambda |s_0|^\rho \text{sign}(s), & \text{if } |s| > s_0, \\ -\lambda |s|^\rho \text{sign}(s), & \text{otherwise}, \end{cases} \tag{2.27}$$

where α, λ are positive constants and $\rho \in (0, 1)$. The sufficient conditions for the finite time convergence to the sliding manifold are

$$\alpha > \frac{C}{K_m}, \quad \lambda^2 \geq \frac{4C}{K_m^2} \frac{K_M(\alpha + C)}{K_m(\alpha - C)}. \tag{2.28}$$

The STA does not need the evaluation of the sign of the time derivative of the sliding variable. For the choice $\rho = 1$, the origin is an exponentially stable equilibrium point. The choice $\rho = 0.5$ assures that the maximum real second order sliding is achieved. For $0 < \rho < 0.5$ the convergence to the origin is even faster. The choice $0 < \rho < 1$ assures the finite time convergence to the origin [25, 29].

Twisting Control Algorithm

This algorithm is characterized by a twisting around the origin, shown in Fig. 2.3. The finite time convergence to the origin of the plane is due to the switching of the control amplitude between two different values. The control amplitude switch at each axis crossing which requires the sign of the time derivative of the sliding variable $\dot{s}$.

In case the relative degree $r = 1$. Consider the system (2.23), the twisting algorithm is defined by the following control law [25]

Fig. 2.3 Twisting algorithm
phase trajectory

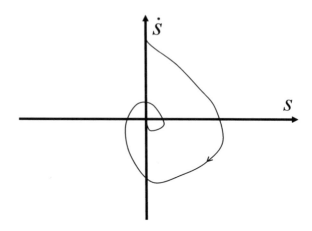

$$\dot{u}(t) = \begin{cases} -u, & \text{if } |u| > 1 \\ -\alpha_m \text{sign}(s), & \text{if } s\dot{s} \leq 0, \ |u| \leq 1 \\ -\alpha_M \text{sign}(s), & \text{if } s\dot{s} > 0, \ |u| \leq 1 \end{cases} \tag{2.29}$$

where $\alpha_M > \alpha_m > 0$ and the sufficient conditions for the finite time convergence to
the sliding manifold are

$$\alpha_m > \frac{4K_M}{s_0}, \quad \alpha_m > \frac{C}{K_m}, \quad K_m \alpha_M > K_M \alpha_m + 2C. \tag{2.30}$$

In case the relative degree $r = 2$. Consider the system (2.26), the twisting algorithm is defined by the following control law [27]

$$u = -r_1 \text{sign}(s) - r_2 \text{sign}(\dot{s}), \ r_1 > r_2 > 0, \tag{2.31}$$

the sufficient conditions for the finite time convergence to the sliding manifold are

$$(r_1 + r_2)K_m > (r_1 - r_2)K_M + 2C, \quad (r_1 - r_2)K_m > C. \tag{2.32}$$

A particular case of the controller with prescribed convergence law [25, 31] is
given by

$$u = -\alpha \text{sign}\left(\dot{s} + \lambda |s|^{\frac{1}{2}} \text{sign}(s)\right), \ \alpha > 0, \ \lambda > 0 \text{ and } \alpha K_m - C > \frac{\lambda^2}{2}. \tag{2.33}$$

Controller (2.33) is close to a terminal sliding mode controller [39].

Sub-Optimal Control Algorithm

The SOSM controller was developed as a sub-optimal feedback implementation
of the classical time optimal control for a double integrator [30]. This algorithm

Fig. 2.4 Sub-optimal
algorithm phase trajectory

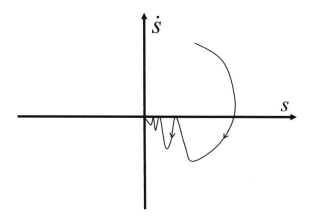

ensures the finite time convergence of s and $\dot{s}$ to zero, confining the trajectories within limit parabolic arcs (including the origin). Both twisting and jumping (in which s and $\dot{s}$ do not change sign) behaviors are possible (see Fig. 2.4). Unlike most SOSMC algorithms, sub-optimal control does not require continuous estimate of $\dot{s}$, only depends on upon the instances when the value of $\dot{s}$ is zero.

Let the relative degree $r = 2$. Consider the system (2.26), the control algorithm is defined by the following control law [2, 25]

$$u(t) = -\alpha(t) V_M \text{sign}(s - \frac{s^*}{2}), \tag{2.34}$$

$$\alpha(t) = \begin{cases} \alpha^*, & \text{if } \left(s - \frac{1}{2}s^*\right)(s^* - s) > 0, \\ 1, & \text{otherwise,} \end{cases}$$

where s^* is the latter singular value of the function s(t) which corresponds to the zero value of $\dot{s}$ and α^* is a positive constant. The sufficient conditions for the finite time convergence to the sliding manifold are

$$\alpha^* \in (0, 1] \cap \left(0, \frac{3K_m}{K_M}\right), \tag{2.35}$$

$$V_M > \max\left(\frac{C}{\alpha^* K_m}, \frac{4C}{3K_m - \alpha^* K_M}\right).$$

2.4 Conclusion

In this chapter, we recalled the basic concepts of classical SMC and SOSM control for uncertain nonlinear systems. Three most common SOSM controllers, i.e., the super-twisting controller, the twisting controller and the sub-optimal controller, are introduced. This control method has proven to be an effective robust control strategy for uncertain nonlinear systems due to its attractive characteristics such as finite time convergence, insensitivity to parameter variations, and complete rejection of external disturbances. SOSM controllers can not only reduce the chattering effect but also improve the accuracy of tracking performance, make this technique suitable for implementation of practical systems.

In the subsequent chapters, we shall concentrate on the applications of sliding mode technique to the observer design and FDI. Then, SMC and FDI of a complex nonlinear system, i.e., PEMFC will be presented with the objective of increasing its efficiency and service life.

References

1. Acary, V., Brogliato, B., Orlov, Y.V.: Chattering-free digital sliding-mode control with state observer and disturbance rejection. IEEE Trans. Autom. Control **57**(5), 1087–1101 (2012)
2. Bartolini, G., Ferrara, A., Usai, E.: Applications of a sub-optimal discontinuous control algorithm for uncertain second order systems. Int. J. Robust Nonlinear Control **7**(4), 299–319 (1997)
3. Bartolini, G., Ferrara, A., Usai, E., Utkin, V.I.: On multi-input chattering-free second-order sliding mode control. IEEE Trans. Autom. Control **45**(9), 1711–1717 (2000)
4. Bartolini, G., Ferrara, A., Usani, E.: Chattering avoidance by second-order sliding mode control. IEEE Trans. Autom. Control **43**(2), 241–246 (1998)
5. Cao, W., Xu, J.: Nonlinear integral-type sliding surface for both matched and unmatched uncertain systems. IEEE Trans. Autom. Control **49**(8), 1355–1360 (2004)
6. Castaños, F., Fridman, L.: Analysis and design of integral sliding manifolds for systems with unmatched perturbations. IEEE Trans. Autom. Control **51**(5), 853–858 (2006)
7. Chan, M., Tao, C., Lee, T.: Sliding mode controller for linear systems with mismatched time-varying uncertainties. J. Frankl. Inst. **337**(2), 105–115 (2000)
8. Choi, H.H.: On the existence of linear sliding surfaces for a class of uncertain dynamic systems with mismatched uncertainties. Automatica **35**(10), 1707–1715 (1999)
9. Choi, H.H.: Variable structure control of dynamical systems with mismatched norm-bounded uncertainties: an LMI approach. Int. J. Control **74**(13), 1324–1334 (2001)
10. Choi, H.H.: LMI-based sliding surface design for integral sliding mode control of mismatched uncertain systems. IEEE Trans. Autom. Control **52**(4), 736–742 (2007)
11. Edwards, C., Spurgeon, S.: Sliding Mode Control: Theory and Applications. CRC Press (1998)
12. Edwards, C., Spurgeon, S.: Sliding Mode Control: Theory and Applications. Taylor & Francis (1998)
13. Efe, M.Ö., Ünsal, C., Kaynak, O., Yu, X.: Variable structure control of a class of uncertain systems. IEEE Trans. Autom. Control **40**(1), 59–64 (2004)
14. Emel'Yanov, S., Korovin, S., Levant, A.: Highorder sliding modes in control systems. Comput. Math. Model. **7**(3), 294–318 (1996)
15. Emelyanov, S., Utkin, V.: Application of automatic control systems of variable structure for the control of objects whose parameters vary over a wide range. DDoklady Akad. Nauk. SSSR **152**(2), 299–301 (1963)
16. Filippov, A., Arscott, F.: Differential Equations with Discontinuous Righthand Sides: Control Systems, vol. 18. Springer (1988)
17. Gao, W., Wang, Y., Homaifa, A.: Discrete-time variable structure control systems. IEEE Trans. Ind. Electron. **42**(2), 117–122 (1995)

18. Gonzalez, T., Moreno, J.A., Fridman, L.: Variable gain super-twisting sliding mode control. IEEE Trans. Autom. Control **57**(8), 2100–2105 (2012)
19. Hung, J.Y., Gao, W., Hung, J.C.: Variable structure control: a survey. IEEE Trans. Ind. Electron. **40**(1), 2–22 (1993)
20. Isidori, A.: Nonlinear Control Systems, vol. 1. Springer (1995)
21. Itkis, U.: Control Systems of Variable Structure. Wiley, New York (1976)
22. Kaynak, O., Erbatur, K., Ertugnrl, M.: The fusion of computationally intelligent methodologies and sliding-mode control–a survey. IEEE Trans. Ind. Electron. **48**(1), 4–17 (2001)
23. Kwan, C.M.: Sliding mode control of linear systems with mismatched uncertainties. IEEE Trans. Autom. Control **31**(2), 303–307 (1995)
24. Lee, H., Utkin, V.I.: Chattering suppression methods in sliding mode control systems. Annu. Rev. Control **31**(2), 179–188 (2007)
25. Levant, A.: Sliding order and sliding accuracy in sliding mode control. Int. J. Control **58**(6), 1247–1263 (1993)
26. Levant, A.: Robust exact differentiation via sliding mode technique. Automatica **34**(3), 379–384 (1998)
27. Levant, A.: Higher-order sliding modes, differentiation and output-feedback control. Int. J. Control **76**(9–10), 924–941 (2003)
28. Levant, A.: Homogeneity approach to high-order sliding mode design. Automatica **41**(5), 823–830 (2005)
29. Moreno, J.A.: Lyapunov analysis of non homogeneous super-twisting algorithms. In: 11th International Workshop on Variable Structure Systems (VSS), pp. 534–539. IEEE (2010)
30. Perruquetti, W., Barbot, J.P.: Sliding Mode Control in Engineering. CRC Press (2002)
31. Sabanovic, A., Fridman, L.M., Spurgeon, S.K.: Variable Structure Systems: From Principles to Implementation, vol. 66. IET (2004)
32. Sellami, A., Arzelier, D., M'hiri, R., Zrida, J.: A sliding mode control approach for systems subjected to a norm-bounded uncertainty. Int. J. Robust Nonlinear Control **17**(4), 327–346 (2007)
33. Slotine, J.J., Sastry, S.S.: Tracking control of nonlinear systems using sliding surface with application to robot manipulator. Int. J. Control **38**(2), 931–938 (1983)
34. Takahashi, R.H.C., Peres, P.L.D.: $\mathcal{H}_2$ guaranteed cost-switching surface design for sliding modes with nonmatching disturbances. IEEE Trans. Autom. Control **44**(11), 2214–2218 (1999)
35. Utkin, V.: Variable structure systems with sliding modes. IEEE Trans. Autom. Control **22**(2), 212–222 (1977)
36. Utkin, V., Gulder, J., Shi, J.: Sliding Mode Control in Electro-mechanical Systems. Automation and Control Engineering Series, vol. 34. Taylor & Francis Group (2009)
37. Wong, L.K., Leung, F.H.F., Tam, P.K.S.: A chattering elimination algorithm for sliding mode control of uncertain non-linear systems. Mechatronics **8**(7), 765–775 (1998)
38. Wu, L., Shi, P., Su, X.: Sliding Mode Control of Uncertain Parameter-switching Hybrid Systems. Wiley (2014)
39. Zhihong, M., Paplinski, A., Wu, H.: A robust MIMO terminal sliding mode control scheme for rigid robotic manipulators. IEEE Trans. Autom. Control **39**(12), 2464–2469 (1994)

Chapter 3
Sliding Mode Observer and Its Applications

SMOs have found wide application in the areas of fault detection, fault reconstruction and health monitoring in recent years. Their well-known advantages are robustness and insensitivity to external disturbance. Higher-order SMOs have better performance as compared to classical sliding mode based observers because their output is continuous and does not require filtering. However, insofar as we are aware, their application in FDI has remained unstudied. In this chapter, we shall develop the theoretical background of SMOs and SMO-based FDI. A bibliographical study of existing approaches in these fields will be followed by a brief presentation of some established first order and second order SMO algorithms. Then, first order SMO based FDI methods will be demonstrated. Finally, our contribution in bridging the gap of second order SMO and adaptive second order SMO based FDI will be presented, followed by two illustrating examples.

3.1 State of the Art

System observation is essential for obtaining unmeasurable states for precise control applications. In general, they are also used for cutting costs by replacing some physical sensors for observable states. Yet, modeling inaccuracy and parametric uncertainty in complex physical systems hinder correct state observation and induce errors. Sliding mode technique is known for its insensitivity to external disturbances, high accuracy and finite time convergence. These properties make it an excellent choice for observation of higher order nonlinear systems such as fuel cell systems.

The early works were based on the assumption that the system under consideration is linear and that a sufficiently accurate mathematical model of the system is available. When the system under consideration is subject to unknown disturbances, the fault signal and unknown disturbance are very likely to produce a similar residual signal. This problem is known in the literature as robust FDI [6], which usually involves two steps: the first step is to decouple the faults of interest from uncertainties and the

© Springer Nature Switzerland AG 2020

J. Liu et al., *Sliding Mode Control Methodology in the Applications of Industrial Power Systems*, Studies in Systems, Decision and Control 249, https://doi.org/10.1007/978-3-030-30655-7_3

second step is to generate residual signals and detect faults by decision logics. Several practical techniques for these steps have been proposed in contemporary literature, for example geometric approaches [22], H_∞-optimization technique [23], observer based approaches (e.g. adaptive observers [35, 38], high gain observers (HGO) [2, 33], unknown input observers (UIO) [3, 13, 26]).

Edwards et al. [6] proposed a fault reconstruction approach based on equivalent output error injection. In this method, the resulting residual signal can approximate the actuator fault to any required accuracy. Based on the work of [6], Tan and Edwards [30] proposed a sensor fault reconstruction method for well-modeled linear systems through the linear matrix inequality (LMI) technique. This approach is of less practical interest, as there is no explicit consideration of disturbance or uncertainty. To overcome this, the same authors [31] proposed an FDI scheme for a class of linear systems with uncertainty, using LMI for minimizing the L_2 gain between the uncertainty and the fault reconstruction signal. Linear uncertain system models can cover a small class of nonlinear systems by representing nonlinear parts as unknown inputs.

Jiang et al. [14] proposed an SMO based fault estimation approach for a class of nonlinear systems with uncertainties. Yan and Edwards [36] proposed a precise fault reconstruction scheme, based on equivalent output error injection, for a class of nonlinear systems with uncertainty. A sufficient condition based on LMI was presented for the existence and stability of a robust SMO, based on strong structural condition of the distribution associated with uncertainties. In their later work [37], this structural constraint was relaxed and the fault distribution vector and the structure matrix of the uncertainty are allowed to be functions of the system's output and input. All these works require that the bounds of the uncertainties and/or faults are *known*.

3.2 First-Order SMO

In this section, several design methods of traditional SMO will be recalled.

3.2.1 SMO Design Based on Utkin's Method

Consider initially the following linear uncertain system [32]:

$$\begin{aligned}\dot{x}(t) &= Ax(t) + Bu(t) + Gd(x, u, t),\\ y(t) &= Cx(t),\end{aligned} \tag{3.1}$$

where $x \in \mathbb{R}^n$ is the state, $u \in \mathbb{R}^m$ is the control input, $y \in \mathbb{R}^p$ is the measurable output. The matrices A, B and C are of appropriate dimensions. It is assumed that $d(x, u, t)$ is unknown, but bounded

$$\|d(x, u, t)\| \le \rho, \quad \forall t \ge 0, \tag{3.2}$$

where $\|\cdot\|$ represents the Euclidean norm. $Gd(x, u, t)$ represents the system uncertainties, with G is a full rank matrix in $\mathbb{R}^{n \times q}$. The matrices B and C are assumed to be of full rank and the pair (A, C) is observable. Furthermore, without loss of generality, the output distribution matrix C can be written as $C = \begin{bmatrix} C_1 & C_2 \end{bmatrix}$, where $C_1 \in \mathbb{R}^{p \times (n-p)}$, $C_2 \in \mathbb{R}^{p \times p}$ and $\det(C_2) \neq 0$. The objective is to estimate the states $x(t)$ only from the measurements of input $u(t)$ and output $y(t)$.

Two cases are considered: $d(x, u, t) = 0$ and $d(x, u, t) \neq 0$. For the first case, a coordinate transformation is introduced in order to facilitate the observer design

$$\begin{bmatrix} x_1(t) \\ y(t) \end{bmatrix} = \begin{bmatrix} I_{n-o} & 0 \\ C_1 & C_2 \end{bmatrix} x = Tx, \tag{3.3}$$

where T is non-singular. With the transformation (3.3), the system (3.1) can be written as

$$\begin{aligned} \dot{x}_1(t) &= A_{11}x_1(t) + A_{12}y(t) + B_1 u(t), \\ \dot{y}(t) &= A_{21}x_1(t) + A_{22}y(t) + B_2 u(t), \end{aligned} \tag{3.4}$$

where

$$TAT^{-1} = \begin{bmatrix} A_{11} & A_{12} \\ A_{21} & A_{22} \end{bmatrix}, \quad TB = \begin{bmatrix} B_1 \\ B_2 \end{bmatrix}, \quad CT^{-1} = \begin{bmatrix} 0 \\ I_p \end{bmatrix}. \tag{3.5}$$

The proposed observer has the following form

$$\begin{aligned} \dot{\hat{x}}_1(t) &= A_{11}\hat{x}_1(t) + A_{12}y(t) + B_1 u(t) + Lv, \\ \dot{\hat{y}}(t) &= A_{21}\hat{x}_1(t) + A_{22}y(t) + B_2 u(t) + v, \end{aligned} \tag{3.6}$$

where $\hat{x}_1(t), \hat{y}(t)$ represent the estimates for $x_1(t), y(t), L \in \mathbb{R}^{(n-p) \times p}$ is the constant gain matrix and the discontinuous terms v is defined

$$v_i = \varrho \text{sign} \left(y_i(t) - \hat{y}_i(t) \right), \quad i = 1, \cdots, p \tag{3.7}$$

where v_i is the ith component of v.

Denote the errors $e_1(t) = x_1(t) - \hat{x}_1(t)$ and $e_y(t) = y(t) - \hat{y}(t)$. Then, the error dynamical system is obtained

$$\dot{e}_1(t) = A_{11}e_1(t) - Lv, \tag{3.8}$$
$$\dot{e}_y(t) = A_{21}e_1(t) - v. \tag{3.9}$$

Thus, for a large enough scalar ϱ, an ideal sliding motion is induced in finite time on the surface

$$S = \{e \in \mathbb{R}^n : e_y = Ce = 0\}. \tag{3.10}$$

During the sliding motion $e_y = \dot{e}_y = 0$, Eq. (3.9) is written as

$$\nu_{eq} = A_{21}e_1(t), \tag{3.11}$$

where ν_{eq} represents the equivalent output error injection term which is generated by a low pass filter. Substituting (3.11) into Eq. (3.8), it follows that the reduced order sliding motion is governed by

$$\dot{e}_1(t) = (A_{11} - LA_{21})\, e_1(t). \tag{3.12}$$

Since the pair (A, C) is observable, then the pair (A_{11}, A_{21}) is also observable. Therefore, there exists L such that the matrix $A_{11} - LA_{21}$ is stable. Consequently, $\hat{x}_1(t)$ converges to $x_1(t)$ asymptotically.

The main disadvantage of the above formulation is the requirement of large value ϱ, to ensure sliding mode for a broad range of initial state estimation errors, especially when the underlying system in unstable. A trade-off between the requirement of a large ϱ and its subsequent reduction to prevent excessive chattering (whilst still ensuring sliding mode) is usually taken into account. Slotine et al. [27] proposed a method which include a linear output error injection term

$$\dot{\hat{y}}(t) = A_{21}\hat{x}_1(t) + A_{22}y(t) + B_2 u(t) + G_l e_y(t) + \nu, \tag{3.13}$$

where the linear gain G_l should be chosen to enhance the size of the so-called *sliding patch*, i.e., the domain of the state estimation error in which sliding occurs. Under certain conditions, the properties of global convergence of state estimation error and robustness can be achieved.

For the second case $d(x, u, t) \neq 0$. Using the transformation (3.3), the linear uncertain system (3.1) can be transformed into following canonical form,

$$\begin{aligned} \dot{x}_1(t) &= A_{11}x_1(t) + A_{12}y(t) + B_1 u(t) + G_1 d(x, u, t), \\ \dot{y}(t) &= A_{21}x_1(t) + A_{22}y(t) + B_2 u(t) + G_2 d(x, u, t). \end{aligned} \tag{3.14}$$

Under the SMO (3.6), the equivalent control signal will be

$$\nu_{eq} = A_{21}e_1(t) + G_2 d(x, u, t). \tag{3.15}$$

The error dynamics of e_1 will become

$$\dot{e}_1(t) = (A_{11} - LA_{21})\, e_1(t) + (G_1 - LG_2)\, d(x, u, t). \tag{3.16}$$

It is clear that $e_1(t)$ will not approach zero if $d(x, u, t)$ is nonzero. It should be noted that even if G_1 is zero, the equivalent control signal will still introduce the uncertainties into its observer error dynamics. A direct approach is to select the gain L such that $G_1 - LG_2 = 0$. However, it may be very difficult to satisfy this condition. Another more reasonable approach is to force the estimation error to be below an acceptable threshold. However, it requires that $\|d(x, u, t)\|$ is small enough, as discussed in [27, 35].

3.2.2 SMO Design Based on Lyapunov Method

The SMO proposed by Wallcott and Żak [34] attempts to provide exponentially convergent estimate of the state described in (3.1) in the presence of *matched* uncertainty.

Recall that the pair (A, C) is assumed to be observable, thus, there exists a matrix K such that $A_0 = A - KC$ is Hurwitz. Therefore, for every real Symmetrical Positive Definite (SPD) matrix $Q \in \mathbb{R}^{n \times n}$, there exists a real SPD matrix P as the unique solution to the following Lyapunov equation

$$A_0^T P + P A_0 = -Q. \tag{3.17}$$

It is also assumed that the structural constraint

$$PG = (FC)^T, \tag{3.18}$$

is satisfied for some $F \in \mathbb{R}^{q \times p}$.

The observer proposed by [34] has the form

$$\dot{\hat{x}}(t) = A\hat{x}(t) + Bu(t) + K(y - C\hat{x}) + \nu, \tag{3.19}$$

where

$$\nu = \begin{cases} \rho(t, y, u)\frac{FCe(t)}{\|FCe(t)\|}, & \text{if } FCe(t) \neq 0, \\ 0, & \text{otherwise.} \end{cases} \tag{3.20}$$

Denote $e(t) = x(t) - \hat{x}(t)$, then, the following error dynamical system is obtained:

$$\dot{e}(t) = A_0 e(t) - \nu + Gd(x, u, t). \tag{3.21}$$

It can directly proof the stability by using $V(e) = e^T P e$ as a Lyapunov function candidate, it is shown that $\dot{V}(e) \leq -cV$, for some positive value c, thus $e(t)$ converges to zero exponentially. Furthermore, an ideal sliding motion takes place on

$$S_F = \{e(t) \in \mathbb{R}^n : FCe(t) = 0\}, \tag{3.22}$$

in finite time.

Remark 3.1. It should be noted that in case of $p > m$, the sliding on S_F is not the same as sliding on S in (3.10). Therefore, the structure of observer (3.19) is different from that in (3.13).

The main difficulty in designing the above observer is the computation the matrices P and F such that (3.17) and (3.18) are satisfied. In [34], a symbolic manipulation tool was used to solve a sequence of constraints that ensure that the principal minors of both P and the right hand side of (3.17) are positive and negative, respectively. It is convenient for low order systems, but not for high order systems. The conditions of the structural requirements (3.17) and (3.18) to be solvable were given by Edwards et al. [6] as follows:

- rank $(CG) = m$;
- any invariant zeros of (A, G, C) lie in the left half plane.

For a square system $(p = m)$, the above two conditions fundamentally require the triple (A, G, C) to be relative degree one and minimum phase. A key development in [6] is that there in no requirement for the pair (A, C) to be observable. The SMOs can be designed as long as the triple (A, G, C) satisfy the above two conditions. Details of the constructive design algorithms can be found in [6, 29]. Floquet and Barbot [9] and Floquet et al. [10] show that the relative degree condition can be relaxed if a classical SMO is combined with the sliding mode robust exact differentiators [17]. Additional independent output signals can be generated from the available measurements.

3.2.3 SMO Design Based on Slotine's Method

Let us consider a nonlinear system in companion form [28]

$$
\begin{aligned}
\dot{x}_1 &= x_2, \\
\dot{x}_2 &= x_3, \\
&\vdots \\
\dot{x}_n &= f(x, t),
\end{aligned}
\tag{3.23}
$$

where $x^{\mathrm{T}} = \begin{bmatrix} x_1 & x_2 & \cdots & x_n \end{bmatrix}$ is the state vector, x_1 is a single measurable output and $f(x, t)$ is a nonlinear uncertain function.

An SMO is designed as follows

$$
\begin{aligned}
\dot{\hat{x}}_1 &= -\alpha_1 e_1 + \hat{x}_2 - k_1 \mathrm{sign}(e_1), \\
\dot{\hat{x}}_2 &= -\alpha_2 e_1 + \hat{x}_3 - k_2 \mathrm{sign}(e_1), \\
&\vdots \\
\dot{\hat{x}}_n &= -\alpha_n e_1 + \hat{f} - k_n \mathrm{sign}(e_1),
\end{aligned}
\tag{3.24}
$$

where $e_1 = \hat{x}_1 - x_1$, $\hat{f}$ is an estimate of $f(x, t)$ and the constants $\alpha_i, i \in \{1, 2, \cdots, n\}$ are chosen to ensure asymptotic convergence for a classical Luenberger observer when $k_i = 0$. The corresponding error dynamics are given by

$$
\begin{aligned}
\dot{e}_1 &= -\alpha_1 e_1 + e_2 - k_1 \text{sign}(e_1), \\
\dot{e}_2 &= -\alpha_2 e_1 + e_3 - k_2 \text{sign}(e_1),
\end{aligned}
$$

$$
\vdots
$$

$$
\dot{e}_n = -\alpha_n e_1 + \tilde{f} - k_n \text{sign}(e_1), \tag{3.25}
$$

where $\tilde{f} = \hat{f} - f(x, t)$ is assumed to be bounded and the following condition holds

$$
k_n \geq |\tilde{f}|. \tag{3.26}
$$

The sliding condition $\frac{d}{dt}(e_1)^2 < 0$ is satisfied in the region

$$
\begin{aligned}
e_2 &\leq k_1 + \alpha_1 e_1, \quad \text{if } e_1 > 0, \\
e_2 &\geq -k_1 + \alpha_1 e_1, \quad \text{if } e_1 < 0.
\end{aligned} \tag{3.27}
$$

Therefore, the sliding mode is attained on $e_1 = \dot{e}_1 = 0$, it follows from Eq. (3.25) that

$$
e_2 - k_1 \text{sign}(e_1) = 0. \tag{3.28}
$$

Substituting (3.28) into (3.25), it follows that the reduced order sliding motion is governed by

$$
\dot{e}_2 = e_3 - \frac{k_2}{k_1} e_2,
$$

$$
\vdots
$$

$$
\dot{e}_n = \tilde{f} - \frac{k_n}{k_1} e_2. \tag{3.29}
$$

The dynamics of *sliding patch* (3.29) are determined by

$$
\left| \lambda I_{n-1} - \begin{pmatrix} -\frac{k_2}{k_1} & 1 & 0 & \cdots & 0 \\ -\frac{k_3}{k_1} & 0 & 1 & \cdots & 0 \\ \vdots & \vdots & \vdots & \vdots & \vdots \\ -\frac{k_n}{k_1} & 0 & 0 & \cdots & 1 \end{pmatrix} \right| = 0. \tag{3.30}
$$

Assuming that k_i, $i \in \{2, \cdots, n\}$ are proportional with k_1 and the poles determining the dynamics of *sliding patch* are critically damped, i.e., are real and equal to some constant values $-\gamma < 0$, then Slotine et al. [28] show the precision of the state estimation error

$$\left| e_2^{(i)} \right| \leq (2\gamma)^i k_1, \quad i = 0, \cdots, n - 2. \tag{3.31}$$

The effect of measurement noise on SMOs was also discussed in [28], the system does not attain a sliding mode in the presence of noise, but remains within a region of the *sliding patch* which is determined by the bound of the noise. Moreover, it was demonstrated that the average dynamics can be modified by the choice of k_i which in turn can tailor the effect of the noise on the state estimates.

3.3 SOSM Observer

It should be noted that the traditional first-order SMOs require low pass filters to obtain equivalent output injections. However, the approximation of the equivalent injections by low pass filters will typically introduce some delays that lead to inaccurate estimates or even to instability for high order systems [9]. To overcome this problem, continuous SOSM algorithms are used to replace the discontinuous first-order sliding mode, such that continuous equivalent output injection signals are obtained. In the following, three kinds of SOSM algorithms will be introduced.

3.3.1 Super-Twisting Algorithm

The STA is described by the differential equation

$$\begin{aligned}
\dot{\tilde{x}}_1 &= -\lambda |\tilde{x}_1|^{\frac{1}{2}} \operatorname{sign}(\tilde{x}_1) + \tilde{x}_2, \\
\dot{\tilde{x}}_2 &= -\alpha \operatorname{sign}(\tilde{x}_1) + \phi(\tilde{x}),
\end{aligned} \tag{3.32}$$

where $\tilde{x} = \begin{bmatrix} \tilde{x}_1 & \tilde{x}_2 \end{bmatrix}^{\mathrm{T}} \in \mathbb{R}^2$ are state variables, λ, α are gains to be designed and the function $\phi(\tilde{x})$ is considered as a perturbation term, which is bounded

$$|\phi(\tilde{x})| \leq \Phi, \tag{3.33}$$

where Φ is a positive constant which is assumed to be *known*. The solutions of (3.32) are all trajectories in the sense of Filippov and Arscott [8].

Frequency Characteristics of STA

It has been shown in [16] that the frequency characteristics of STA can be used to evaluate the performance of STA observer and determine the choice of its parameters.

Let $\tilde{x}_1 = A \sin(\omega t)$ and use the describing function method, we have

$$
\begin{aligned}
\frac{1}{\pi} \int_0^{2\pi} |\tilde{x}_1|^{\frac{1}{2}} \, \text{sign}(\tilde{x}_1) \sin(\omega\tau) d\omega\tau \\
= \frac{2}{\pi} \int_0^{\pi} |A \sin(\omega\tau)|^{\frac{1}{2}} \, \text{sign}\,(A \sin(\omega\tau)) \sin(\omega\tau) d\omega\tau \\
= \frac{2}{\pi} A^{\frac{1}{2}} \int_0^{\pi} |\sin(\omega\tau)|^{\frac{3}{2}} d\omega\tau,
\end{aligned}
\tag{3.34}
$$

$$
\frac{1}{\pi} \int_0^{2\pi} \text{sign}(\tilde{x}_1) \sin(\omega\tau) d\omega\tau = \frac{2}{\pi} \int_0^{\pi} \text{sign}\,(A \sin(\omega\tau)) \sin(\omega\tau) d\omega\tau = \frac{4}{\pi}. \tag{3.35}
$$

It is easy to get $\frac{2}{\pi} \int_0^{\pi} |\sin(\omega\tau)| \, d\omega\tau = \frac{4}{\pi}$, $\frac{2}{\pi} \int_0^{\pi} |\sin(\omega\tau)|^2 \, d\omega\tau = 1$ and

$$
1 \leq \Delta = \frac{2}{\pi} \int_0^{\pi} |\sin(\omega\tau)|^{\frac{3}{2}} \, d\omega\tau \leq \frac{4}{\pi}. \tag{3.36}
$$

Thus, the describing functions of the nonlinear terms $(|\cdot|^{\frac{1}{2}} \text{sign}(\cdot), \text{sign}(\cdot))$ in (3.32) are $N_1(A) = \frac{\Delta}{A^{\frac{1}{2}}}$ and $N_2(A) = \frac{4}{\pi A}$, respectively.

The linearization of system (3.32) is

$$
\begin{aligned}
\dot{\tilde{x}}_1 &= -\lambda \Delta A^{-\frac{1}{2}} \tilde{x}_1 + \tilde{x}_2, \\
\dot{\tilde{x}}_2 &= -\alpha \frac{4}{\pi A} \tilde{x}_1 + \phi(\tilde{x}).
\end{aligned}
\tag{3.37}
$$

The nature frequency and its damping coefficient of system (3.37) are

$$
\omega_n = \frac{2\sqrt{\alpha}}{\sqrt{\pi} A^{\frac{1}{2}}}, \quad \varsigma = \Delta \frac{\lambda\sqrt{\pi}}{4\sqrt{\alpha}}. \tag{3.38}
$$

From (3.37) and (3.38), we can find that, when the observation error $\tilde{x}_1$ is large, that is A is large far from the origin, the gains in (3.37) are small. On the other hand, when the observation error $\tilde{x}_1$ is small, that is A is small near the origin, the gains are large. Therefore, we can conclude that the behavior of STA is strong near the origin, contrarily, is weak far from the origin. The frequency characteristic of STA is shown in Fig. 3.1.

Convergence of STA

A strong Lyapunov function was proposed to analyze the stability and finite time convergence of the STA [19]

$$
V(\zeta) = \zeta^{\mathrm{T}} P \zeta, \tag{3.39}
$$

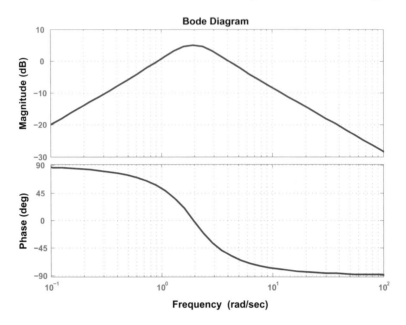

Fig. 3.1 Frequency characteristics of STA

where $\zeta^{\mathrm{T}} = [\zeta_1, \zeta_2] = \left[|\tilde{x}_1|^{\frac{1}{2}} \operatorname{sign}(\tilde{x}_1), \tilde{x}_2 \right]$ and $P = \frac{1}{2} \begin{bmatrix} 4\alpha + \lambda^2 & -\lambda \\ -\lambda & 2 \end{bmatrix}$ is a SPD matrix. The derivative of the vector ζ is given by

$$\dot{\zeta} = \frac{1}{|\zeta_1|} A\zeta + B\phi(\tilde{x}), \quad A = \begin{bmatrix} -\frac{1}{2}\lambda & \frac{1}{2} \\ -\alpha & 0 \end{bmatrix},$$

$$B = \begin{bmatrix} 0 & 1 \end{bmatrix}^{\mathrm{T}}, \quad \zeta_1 = \begin{bmatrix} 1 & 0 \end{bmatrix} \zeta = C\zeta. \tag{3.40}$$

It should be noted that $V(\zeta)$ in (3.39) is continuous but not differentiable on the set $S = \{(x_1, x_2) \in \mathcal{R}^2 | x_1 = 0\}$. Thus, the classical versions of Lyapunov's theorem [1, 21] can not be applied since they require a continuously differentiable of at least a locally Lipschitz continuous Lyapunov function. However, by means of Zubov'theorem [41] which requires only continuous Lyapunov functions, it is still possible to show the convergence property [19].

Proposition 3.1 [4, 19]. *Suppose that the condition* (3.33) *is satisfied and there exists a SPD matrix P such that the Algebraic Riccati Equation (ARE)*

$$A^{\mathrm{T}}P + PA + \Phi^2 C^{\mathrm{T}}C + PBB^{\mathrm{T}}P = -Q < 0 \tag{3.41}$$

is satisfied. Then, all trajectories of system (3.32) *converge to the origin in finite time. Moreover, the quadratic form* $V(\zeta) = \zeta^{\mathrm{T}} P \zeta$ *is a* strong, robust Lyapunov *function for the system* (3.32) *and the finite convergence time T is estimated*

$$T = \frac{2}{\gamma(P)} V^{\frac{1}{2}}(\tilde{x}_0),$$ (3.42)

where $\tilde{x}_0$ is the initial error and $\gamma(P)$ is

$$\gamma(P) = \frac{\lambda_{\min}(Q)\lambda_{\min}^{\frac{1}{2}}(P)}{\lambda_{\max}(P)}.$$ (3.43)

Proof of this Proposition 3.1 can be found in [4, 19].

3.3.2 Modified Super-Twisting Algorithm

It is easy to see that the standard (constant gains) STA always requires the condition (3.33). It does not allow to compensate uncertainties/disturbances growing in time or together with the state variables due to its homogeneous nature. This means that the standard STA can not ensure the sliding motions even for systems for which the linear part is not exactly known. Thus, it is very important to design non-homogeneous extension of the standard STA allowing exact compensation of the smooth uncertainties/disturbances bounded together with their derivatives by the known functions, which could grow together with the state [11].

As shown in [20] that linear growing perturbations are included by means of the addition of linear terms to the nonlinear SOSM terms (SOSML). The behavior of the STA near the origin is significantly improved compared with the linear case. Conversely, the additional linear term improves the behavior of the STA when the states are far from the origin. In other words, the linear terms can deal with a bounded perturbation with linear growth in time while the nonlinear terms of STA can deal with a strong perturbation near the origin. Therefore, the SOSML inherits the best properties of both the linear and the nonlinear terms.

The SOSML algorithm is described by the following differential equation

$$\begin{aligned}\dot{\tilde{x}}_1 &= -\lambda |\tilde{x}_1|^{\frac{1}{2}} \operatorname{sign}(\tilde{x}_1) - k_\lambda \tilde{x}_1 + \tilde{x}_2, \\ \dot{\tilde{x}}_2 &= -\alpha \operatorname{sign}(\tilde{x}_1) - k_\alpha \tilde{x}_1 + \phi(\tilde{x}),\end{aligned}$$ (3.44)

where $\lambda, \alpha, k_\lambda, k_\alpha$ are positive gains to be determined and the perturbation term $\phi(\tilde{x})$ is bounded by

$$|\phi(\tilde{x})| \leq \delta_1 + \delta_2 |\tilde{x}_1|,$$ (3.45)

where δ_1 and δ_2 are some positive constant and are assumed to be *known*. The solutions of (3.44) are all trajectories in the sense of Filippov [8].

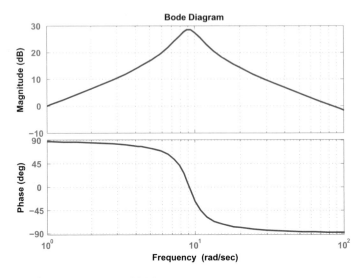

Fig. 3.2 Frequency characteristics of SOSML

Frequency Characteristics of SOSML

Let $\tilde{x}_1 = A_1 \sin(\omega_1 t)$, the linearization of system (3.44) is given as

$$
\begin{aligned}
\dot{\tilde{x}}_1 &= -\left(\lambda \Delta A_1^{-\frac{1}{2}} + k_\lambda\right)\tilde{x}_1 + \tilde{x}_2, \\
\dot{\tilde{x}}_2 &= -\left(\frac{4\alpha}{\pi A_1} + k_\alpha\right)\tilde{x}_1 + \phi(\tilde{x}).
\end{aligned}
\tag{3.46}
$$

The nature frequency and its damping coefficient of system (3.46) are

$$
\omega_n = \sqrt{\frac{4\alpha}{\pi A_1} + k_\alpha}, \quad \varsigma = \frac{\lambda \Delta A_1^{-\frac{1}{2}} + k_\lambda}{2\sqrt{\frac{4\alpha}{\pi A_1} + k_\alpha}}.
\tag{3.47}
$$

Comparing with STA, the behavior of SOSML is both strong when the observation error $\tilde{x}_1$ is near the origin and far from the origin. Better dynamic characteristics are obtained which can be seen from the frequency characteristic of SOSML shown in Fig. 3.2.

Convergence of SOSML

A strong Lyapunov function was proposed to analyze the stability and finite time convergence of the SOSML

$$
V_1(\tilde{x}) = 2\alpha|\tilde{x}_1| + k_\alpha\tilde{x}_1^2 + \frac{1}{2}\tilde{x}_2^2 + \frac{1}{2}(\lambda|\tilde{x}_1|^{1/2}\mathrm{sign}(\tilde{x}_1) + k_\lambda\tilde{x}_1 - \tilde{x}_2)^2.
$$

The Eq. (3.44) can be written as a quadratic form

$$V_1(\tilde{x}) = \xi^{\mathrm{T}} P_1 \xi,$$ (3.48)

where

$$\xi^{\mathrm{T}} = \left[\xi_1, \xi_2, \xi_3\right] = \left[|\tilde{x}_1|^{\frac{1}{2}} \operatorname{sign}(\tilde{x}_1),\ \tilde{x}_1,\ \tilde{x}_2\right],$$

and

$$P_1 = \frac{1}{2} \begin{bmatrix} 4\alpha + \lambda^2 & \lambda k_\lambda & -\lambda \\ \lambda k_\lambda & 2k_\alpha k_\lambda^2 & -k_\lambda \\ -\lambda & -k_\lambda & 2 \end{bmatrix}.$$

The matrix P is SPD under the condition

$$4\alpha k_\alpha > (8\alpha + 9\lambda^2)k_\lambda^2.$$ (3.49)

The derivative of the vector ζ is given by

$$\dot{\xi} = \frac{1}{|\xi_1|} \underbrace{\begin{bmatrix} -\frac{\lambda}{2} & 0 & \frac{1}{2} \\ 0 & -\lambda & 0 \\ -\alpha & 0 & 0 \end{bmatrix}}_{A_1} \xi + \underbrace{\begin{bmatrix} -\frac{k_\lambda}{2} & 0 & 0 \\ 0 & -k_\lambda & 1 \\ 0 & -k_\alpha & 0 \end{bmatrix}}_{A_2} \xi + \begin{bmatrix} 0 \\ 0 \\ 1 \end{bmatrix} \phi(\tilde{x}).$$ (3.50)

Proposition 3.2. *Suppose conditions (3.45) and (3.49) hold. Then, all trajectories of system (3.44) converge to the origin in finite time if*

$$\lambda_{\min}(\Omega_1) > \delta_1 \sqrt{\lambda^2 + k_\lambda^2 + 4},$$
$$\lambda_{\min}(\Omega_2) > \delta_2 \lambda_{\max}(\Delta\Phi),$$ (3.51)

where the matrices Ω_1, Ω_2 and $\Delta\Phi$ are defined in (3.55) and (3.56), respectively. Moreover, the quadratic form $V_1(x) = \xi^{\mathrm{T}} P_1 \xi$ is a strong, robust Lyapunov function for the system (3.44) and the finite convergence time T_1 is estimated

$$T_1 = \frac{2}{\gamma_1} V_1^{\frac{1}{2}}(\tilde{x}_0),$$ (3.52)

where $\tilde{x}_0$ is the initial error and γ_1 is some positive constant.

Proof. The time derivative of Lyapunov function candidate (3.48) along the trajectories of the system (3.50) is calculated as,

$$\dot{V}_1 = \frac{1}{|\xi_1|} \xi^{\mathrm{T}} (A_1^{\mathrm{T}} P_1 + P_1 A_1) \xi + \xi^{\mathrm{T}} (A_2^{\mathrm{T}} P_1 + P_1 A_2) \xi + 2\xi^{\mathrm{T}} P_1 \begin{bmatrix} 0 \\ 0 \\ \phi(\tilde{x}) \end{bmatrix}.$$ (3.53)

It follows from $\xi_1\xi_2 = \frac{\xi_2^2}{|\xi_1|}$ and $\xi_1\xi_3 = \frac{\xi_2\xi_3}{|\xi_1|}$ that

$$\dot{V}_1 = -\frac{1}{|\xi_1|}\xi^T\Omega_1\xi - \xi^T\Omega_2\xi + \delta_1 g_1\xi + \delta_2\xi^T g_1^T \begin{bmatrix} 0 & 1 & 0 \end{bmatrix}\xi, \qquad (3.54)$$

where

$$g_1 = \begin{bmatrix} -\lambda & -k_\lambda & 2 \end{bmatrix},$$

$$\Omega_1 = \frac{\lambda}{2}\begin{bmatrix} \lambda^2 + 2\alpha & 0 & -\lambda \\ 0 & 2k_\alpha + 5k_\lambda^2 & -3k_\lambda \\ -\lambda & -3k_\lambda & 1 \end{bmatrix},$$

$$\Omega_2 = k_\lambda\begin{bmatrix} \alpha + 2\lambda^2 & 0 & 0 \\ 0 & k_\alpha + k_\lambda^2 & -k_\lambda \\ 0 & -k_\lambda & 1 \end{bmatrix}.$$

$$(3.55)$$

It is easy to show that Ω_1, Ω_2 are positive definite under the condition (3.49). Therefore, Eq. (3.54) can be rewritten as,

$$\dot{V}_1 \leq -\frac{1}{|\xi_1|}\lambda_{\min}(\Omega_1)\|\xi\|^2 - \lambda_{\min}(\Omega_2)\|\xi\|^2 + \delta_1\|g_1\|\|\xi\| + \delta_2\lambda_{\max}(\Delta\Phi)\|\xi\|^2,$$

where

$$\Delta\Phi = \begin{bmatrix} \frac{\lambda}{2} & 0 & 0 \\ 0 & \frac{\lambda}{2} - k_\lambda + 1 & 0 \\ 0 & 0 & 1 \end{bmatrix}. \qquad (3.56)$$

Further, using the fact that

$$|\xi_1| \leq \|\xi\| \leq \frac{V_1^{\frac{1}{2}}}{\lambda_{\min}^{\frac{1}{2}}(P_1)}, \qquad (3.57)$$

$$\|g_1\| = \sqrt{\lambda^2 + k_\lambda^2 + 4},$$

it follows that

$$\dot{V}_1 \leq -\left(\lambda_{\min}(\Omega_1) - \delta_1\sqrt{\lambda^2 + k_\lambda^2 + 4}\right)\|\xi\| - (\lambda_{\min}(\Omega_2) - \delta_2\lambda_{\max}(\Delta\Phi))\|\xi\|^2.$$

Under the conditions that $c_1 = \lambda_{\min}(\Omega_1) - \delta_1\sqrt{\lambda^2 + k_\lambda^2 + 4}$ and $c_2 = \lambda_{\min}(\Omega_2) - \delta_2\lambda_{\max}(\Delta\Phi)$ are positive, it follows

$$\dot{V}_1 \leq -\gamma_1 V_1^{\frac{1}{2}} - \gamma_2 V_1, \qquad (3.58)$$

where

$$\gamma_1 = \frac{c_1}{\lambda_{\max}^{\frac{1}{2}}(P_1)}, \quad \gamma_2 = \frac{c_2}{\lambda_{\max}(P_1)}. \tag{3.59}$$

By the comparison lemma [15], it follows that $V(\tilde{x}(t))_1$ and $\tilde{x}(t)$ converge to zero in finite time and reaches that value at most after $T_1 = \frac{2V_1^{\frac{1}{2}}(\tilde{x}_0)}{\gamma_1}$ units of time.

3.3.3   Adaptive-Gain Super-Twisting Algorithm

The main disadvantage of STA and SOSML is that it requires the knowledge of the bound of uncertainty, see conditions (3.33) and (3.45). However, in many practical cases, this boundary can not be easily obtained. Therefore, an adaptive-gain SOSML is proposed, which handles the uncertainty with the *unknown* boundary. An adaptive law of the gains of the proposed algorithm is derived via the so-called "time scaling" approach [25], which are adapted dynamically according to the observation error.

The adaptive-gain SOSML algorithm is described as follows

$$\dot{\tilde{x}}_1 = -\lambda(t)\,|\tilde{x}_1|^{\frac{1}{2}}\,\mathrm{sign}(\tilde{x}_1) - k_\lambda(t)\tilde{x}_1 + \tilde{x}_2,$$
$$\dot{\tilde{x}}_2 = -\alpha(t)\mathrm{sign}(\tilde{x}_1) - k_\alpha(t)\tilde{x}_1 + \phi(\tilde{x}), \tag{3.60}$$

the perturbation term $\phi(\tilde{x})$ is bounded by

$$|\phi(\tilde{x})| \le \sigma_1 + \sigma_2\,|\tilde{x}_1|, \tag{3.61}$$

where σ_1 and σ_2 are some positive constant and are assumed to be *unknown*.

The adaptive gains $\lambda(t), \alpha(t), k_\lambda(t)$ and $k_\alpha(t)$ are formulated as

$$\lambda(t) = \lambda_0\sqrt{l(t)}, \quad \alpha(t) = \alpha_0 l(t),$$
$$k_\lambda(t) = k_{\lambda_0} l(t), \quad k_\alpha(t) = k_{\alpha_0} l^2(t), \tag{3.62}$$

where $\lambda_0, \alpha_0, k_{\lambda_0}$ and k_{α_0} are positive constants to be defined and $l(t)$ is a positive, time-varying, scalar function.

The adaptive law of the time-varying function $l(t)$ is given by:

$$\dot{l}(t) = \begin{cases} k, & \text{if } |\tilde{x}_1| \ne 0, \\ 0, & \text{otherwise}, \end{cases} \tag{3.63}$$

where k is a positive constant.

Theorem 3.1 Consider system (3.60), (3.62), and (3.63). *Suppose that the condition (3.61) hold with some **unknown** constants σ_1 and σ_2. The trajectories of the system (3.60) converge to zero in finite time if the following condition is satisfied*

$$4\alpha_0 k_{\alpha_0} > 8k_{\lambda_0}^2 \alpha_0 + 9\lambda_0^2 k_{\lambda_0}^2. \tag{3.64}$$

Proof. A new state vector is introduced to represent the system (3.60) in a more convenient form for Lyapunov analysis.

$$\zeta = \begin{bmatrix} \zeta_1 \\ \zeta_2 \\ \zeta_3 \end{bmatrix} = \begin{bmatrix} l^{\frac{1}{2}}(t)|\tilde{x}_1|^{\frac{1}{2}}\text{sign}(\tilde{x}_1) \\ l(t)\tilde{x}_1 \\ \tilde{x}_2 \end{bmatrix}. \tag{3.65}$$

Thus, the system in (3.60) can be rewritten as

$$\dot{\zeta} = \frac{l(t)}{|\zeta_1|}\underbrace{\begin{bmatrix} -\frac{\lambda_0}{2} & 0 & \frac{1}{2} \\ 0 & -\lambda_0 & 0 \\ -\alpha_0 & 0 & 0 \end{bmatrix}}_{A_1}\zeta + l(t)\underbrace{\begin{bmatrix} -\frac{k_{\lambda_0}}{2} & 0 & 0 \\ 0 & -k_{\lambda_0} & 1 \\ 0 & -k_{\alpha_0} & 0 \end{bmatrix}}_{A_2}\zeta + \begin{bmatrix} \frac{\dot{l}}{2l(t)}\zeta_1 \\ \frac{\dot{l}}{2l(t)}\zeta_2 \\ \phi(\tilde{x}) \end{bmatrix}. \tag{3.66}$$

Then, the following Lyapunov function candidate is introduced for the system (3.66):

$$V(\zeta) = 2\alpha_0\zeta_1^2 + k_{\alpha_0}\zeta_2^2 + \frac{1}{2}\zeta_3^2 + \frac{1}{2}\left(\lambda_0\zeta_1 + k_{\lambda_0}\zeta_2 - \zeta_3\right)^2, \tag{3.67}$$

which can be rewritten as a quadratic form

$$V(\zeta) = \zeta^{\mathrm{T}}P\zeta, \quad P = \frac{1}{2}\begin{bmatrix} 4\alpha_0 + \lambda_0^2 & \lambda_0 k_{\lambda_0} & -\lambda_0 \\ \lambda_0 k_{\lambda_0} & k_{\lambda_0}^2 + 2k_{\alpha_0} & -k_{\lambda_0} \\ -\lambda_0 & -k_{\lambda_0} & 2 \end{bmatrix}. \tag{3.68}$$

As (3.67) is a continuous Lyapunov function and the matrix P is positive definite. Taking the derivative of (3.67) along the trajectories of (3.66),

$$\dot{V} = -\frac{l(t)}{|\zeta_1|}\zeta^{\mathrm{T}}\Omega_1\zeta - l(t)\zeta^{\mathrm{T}}\Omega_2\zeta + \phi(\tilde{x})q_1\zeta + \frac{\dot{l}(t)}{l(t)}q_2 P\zeta, \tag{3.69}$$

where $q_1 = \begin{bmatrix} -\lambda_0 & -k_{\lambda_0} & 2 \end{bmatrix}$, $q_2 = \begin{bmatrix} \zeta_1 & \zeta_2 & 0 \end{bmatrix}$, and

$$\Omega_1 = \frac{\lambda_0}{2} \begin{bmatrix} \lambda_0^2 + 2\alpha_0 & 0 & -\lambda_0 \\ 0 & 2k_{\alpha_0} + 5k_{\lambda_0}^2 & -3k_{\lambda_0} \\ -\lambda_0 & -3k_{\lambda_0} & 1 \end{bmatrix},$$

$$\Omega_2 = k_{\lambda_0} \begin{bmatrix} \alpha_0 + 2\lambda_0^2 & 0 & 0 \\ 0 & k_{\alpha_0} + k_{\lambda_0}^2 & -k_{\lambda_0} \\ 0 & -k_{\lambda_0} & 1 \end{bmatrix}. \tag{3.70}$$

It is easy to verify that Ω_1 and Ω_2 are positive definite matrices under the condition (3.64).

Using the fact $\lambda_{\min}(P)\|\zeta\|^2 \le V \le \lambda_{\max}(P)\|\zeta\|^2$ and $\|q_1\| = \sqrt{\lambda^2 + k_\lambda^2 + 4}$. Equation (3.69) can be rewritten as

$$\dot{V} \le -l(t)\frac{\lambda_{\min}(\Omega_1)}{\lambda_{\max}^{\frac{1}{2}}(P)}V^{\frac{1}{2}} - l(t)\frac{\lambda_{\min}(\Omega_2)}{\lambda_{\max}(P)}V + \frac{\sigma_1\|q_1\|_2}{\lambda_{\min}^{\frac{1}{2}}(P)}V^{\frac{1}{2}}$$
$$+ \frac{\sigma_2}{l(t)}\zeta^T \Delta\Phi\zeta + \frac{\dot{l}(t)}{2l(t)}\Delta\Omega, \tag{3.71}$$

where

$$\Delta\Omega = \left((4\alpha_0 + \lambda_0^2)\zeta_1^2 + 2\lambda_0 k_{\lambda_0}\zeta_1\zeta_2 + 2k_{\alpha_0}k_{\lambda_0}^2\zeta_2^2 - \lambda_0\zeta_1\zeta_3 - k_{\lambda_0}\zeta_2\zeta_3\right)$$
$$\le \zeta^T Q \zeta,$$

$$\Delta\Phi = \begin{bmatrix} \frac{\lambda}{2} & 0 & 0 \\ 0 & \frac{\lambda}{2} - k_\lambda + 1 & 0 \\ 0 & 0 & 1 \end{bmatrix}, \tag{3.72}$$

$$Q = \begin{bmatrix} 4\alpha_0 + \lambda_0^2 + \lambda_0 k_{\lambda_0} + \frac{\lambda_0}{2} & 0 & 0 \\ 0 & 2k_{\alpha_0}k_{\lambda_0}^2 + \lambda_0 k_{\lambda_0} + \frac{k_{\lambda_0}}{2} & 0 \\ 0 & 0 & \frac{\lambda_0 + k_{\lambda_0}}{2} \end{bmatrix}.$$

In view of (3.72), Eq. (3.71) becomes

$$\dot{V} \le -\left(l(t)\frac{\lambda_{\min}(\Omega_1)}{\lambda_{\max}^{\frac{1}{2}}(P)} - \frac{\sigma_1\sqrt{\lambda^2 + k_\lambda^2 + 4}}{\lambda_{\min}^{\frac{1}{2}}(P)}\right)V^{\frac{1}{2}}$$
$$- \left(l(t)\frac{\lambda_{\min}(\Omega_2)}{\lambda_{\max}(P)} - \frac{\sigma_2}{l(t)}\frac{\lambda_{\max}(\Delta\Phi)}{\lambda_{\min}(P)} - \frac{\dot{l}(t)}{2l(t)}\frac{\lambda_{\max}(Q)}{\lambda_{\min}(P)}\right)V. \tag{3.73}$$

For simplicity, we define

$$\gamma_1 = \frac{\lambda_{\min}(Q)}{\lambda_{\max}^{\frac{1}{2}}(P)}, \quad \gamma_2 = \frac{\sigma_1\sqrt{\lambda^2 + k_\lambda^2 + 4}}{\lambda_{\min}^{\frac{1}{2}}(P)}, \quad \gamma_3 = \frac{\lambda_{\min}(\Omega_2)}{\lambda_{\max}(P)},$$
$$\gamma_4 = \sigma_2\frac{\lambda_{\max}(\Delta\Phi)}{\lambda_{\min}(P)}, \quad \gamma_5 = \frac{\lambda_{\max}(Q)}{2\lambda_{\min}(P)}, \tag{3.74}$$

where $\gamma_1, \gamma_2, \gamma_3, \gamma_4$ and γ_5 are all positive constants. Thus, Eq. (3.73) can be simplified as

$$\dot{V} \le -\left(l(t)\gamma_1 - \gamma_2\right) V^{\frac{1}{2}} - \left(l(t)\gamma_3 - \frac{\gamma_4}{l(t)} - \frac{\dot{l}(t)}{l(t)}\gamma_5\right) V. \qquad (3.75)$$

Because $\dot{l}(t) \ge 0$ such that the terms $l(t)\gamma_1 - \gamma_2$ and $l(t)\gamma_3 - \frac{\gamma_4}{l(t)} - \frac{\dot{l}(t)}{l(t)}\gamma_5$ are positive in finite time, it follows from (3.75) that

$$\dot{V} \le -c_1 V^{\frac{1}{2}} - c_2 V, \qquad (3.76)$$

where c_1 and c_2 are positive constants. By the comparison principle [15], it follows that $V(\zeta)$ and therefore ζ converge to zero in finite time. Thus, Theorem 3.1 is proven.

3.4 SMO-Based FDI

Let us now explore the utilization of SMOs for system fault detection and reconstruction. Most observer-based FDI schemes generate residuals by comparing the measurement and its corresponding estimate provided by observers. Wrong estimates will be produced when faults occur, thus, a nonzero residual would raise an alarm. In this section, it is shown that SMO can not only detect the faults but also reconstruct the fault signal (its shape and magnitude). Fault reconstruction is of great interest in active fault tolerant control which can be employed in controller design [7].

Consider a nonlinear system [36]

$$\dot{x} = Ax + G(x, u) + E\Psi(t, x, u) + Df(y, u, t),$$
$$y = Cx, \qquad (3.77)$$

where $x \in \mathbb{R}^n$, $u \in \mathbb{R}^m$ and $y \in \mathbb{R}^p$ are the state variables, inputs and outputs, respectively. $A \in \mathbb{R}^{n \times n}$, $E \in \mathbb{R}^{n \times r}$, $D \in \mathbb{R}^{n \times q}$ and $C \in \mathbb{R}^{p \times n}$ $(n > p > q)$ are constant matrices. The matrices C and D are assumed to be of full rank. The known nonlinear term $G(x, u)$ is Lipschitz with respect to x uniformly for $u \in U$, where U is an admissible control set. The bounded unknown function $f(y, u, t) \in \mathbb{R}^q$ represents the actuator fault which needs to be estimated and the uncertain nonlinear term $\Psi(t, x, u)$ represents the modeling uncertainties and disturbances affecting the system.

First, some assumptions will be imposed on the system (3.77).

Assumption 3.1. $\text{rank}(C[E, D]) = \text{rank}([E, D]) = \tilde{q} \le p$.

Assumption 3.2. The invariant zeros of the matrix triple $(A, [E, D], C)$ lie in the left half plane.

Assumption 3.3. The function $f(t, u)$ and its time derivative are unknown but bounded:

$$\|f(y, u, t)\| \le \alpha(y, u, t), \quad \|\dot{f}(y, u, t)\| \le \alpha_d(y, u, t), \tag{3.78}$$

where $\alpha(y, u, t)$ and $\alpha_d(y, u, t)$ are two known functions.

Assumption 3.4. The nonlinear term $\Psi(t, x, u)$ and its time derivative are unknown but bounded:

$$\|\Psi(t, x, u)\| \le \beta, \quad \|\dot{\Psi}(t, x, u)\| \le \beta_d, \tag{3.79}$$

where β and β_d are known positive scalars.

Under Assumption 3.1, there exists a coordinate system in which the triple $(A, [E, \dot{D}], C)$ has the following structure

$$\left(\begin{bmatrix} A_1 & A_2 \\ A_3 & A_4 \end{bmatrix}, \begin{bmatrix} 0_{(n-p)\times r} & 0_{(n-p)\times q} \\ E_2 & D_2 \end{bmatrix}, \begin{bmatrix} 0_{p\times(n-p)} & C_2 \end{bmatrix} \right), \tag{3.80}$$

where $A_1 \in \mathbb{R}^{(n-p)\times(n-p)}$, $C_2 \in \mathbb{R}^{p\times p}$ is nonsingular and

$$E_2 = \begin{bmatrix} 0_{(p-\tilde{q})\times r} \\ E_{22} \end{bmatrix}, \quad D_2 = \begin{bmatrix} 0_{(p-\tilde{q})\times q} \\ D_{22} \end{bmatrix}, \tag{3.81}$$

with $E_{22} \in \mathbb{R}^{\tilde{q}\times r}$ and $D_{22} \in \mathbb{R}^{\tilde{q}\times q}$ of full rank.

Under Assumption 3.2, there exists a matrix $L \in \mathbb{R}^{(n-p)\times p}$ with the form

$$L = \begin{bmatrix} L_1 & 0_{(n-p)\times\tilde{q}} \end{bmatrix}, \tag{3.82}$$

with $L_1 \in \mathbb{R}^{(n-p)\times(p-\tilde{q})}$ such that $A_1 + LA_3$ is Hurwitz.

Without loss of generality, the system (3.77) has the form

$$\begin{aligned}
\dot{x}_1 &= A_1 x_1 + A_2 x_2 + G_1(x, u), \\
\dot{x}_2 &= A_3 x_1 + A_4 x_2 + G_2(x, u) + E_2\Psi(t, x, u) + D_2 f(y, u, t), \\
y &= C_2 x_2,
\end{aligned} \tag{3.83}$$

where $x := \text{col}(x_1, x_2)$, $x_1 \in \mathbb{R}^{n-p}$, $x_2 \in \mathbb{R}^p$, $G_1(x, u)$ and $G_2(x, u)$ are the first $n - p$ and the last p components of $G(x, u)$, respectively. The matrices E_2 and D_2 are given in Eq. (3.81).

3.4.1   First-Order SMO-Based FDI

Consider the system (3.83), there exists a coordinate transformation $z = Tx$

$$T := \begin{bmatrix} I_{n-p} & L \\ 0 & I_p \end{bmatrix}, \tag{3.84}$$

where L is defined in (3.82). Thus, in the new coordinate z, the system (3.83) has the following form

$$
\begin{aligned}
\dot{z}_1 &= F_1 z_1 + F_2 z_2 + \begin{bmatrix} I_{n-p} & L \end{bmatrix} G(T^{-1}z, u), \\
\dot{z}_2 &= A_3 z_1 + F_3 z_2 + G_2(T^{-1}z, u) + E_2 \Psi(t, T^{-1}z, u) + D_2 f(y, u, t), \\
y &= C_2 z_2,
\end{aligned}
\tag{3.85}
$$

where $z := \mathrm{col}(z_1, z_2)$, $z_1 \in \mathbb{R}^{n-p}$, $z_2 \in \mathbb{R}^p$ and

$$F_1 = A_1 + L A_3, \quad F_2 = A_2 + L A_4 - F_1 L, \quad F_3 = A_4 - L A_3, \tag{3.86}$$

with the matrix F_1 is Hurwitz.

Consider the following dynamical observer for the system (3.85)

$$
\begin{aligned}
\dot{\hat{z}}_1 &= F_1 \hat{z}_1 + F_2 C_2^{-1} y + \begin{bmatrix} I_{n-p} & L \end{bmatrix} G(T^{-1}\hat{z}, u), \\
\dot{\hat{z}}_2 &= A_3 \hat{z}_1 + F_3 \hat{z}_2 - K(y - C_2 \hat{z}_2) + G_2(T^{-1}\hat{z}, u) + \nu(y, \hat{y}, \hat{z}, u, t), \\
\hat{y} &= C_2 \hat{z}_2,
\end{aligned}
\tag{3.87}
$$

where $\hat{z} := \mathrm{col}(\hat{z}_1, C_2^{-1}y)$ and $\hat{y}$ is the output of the observer system. The gain matrix K is chosen such that

$$F := C_2 F_3 C_2^{-1} + C_2 K, \tag{3.88}$$

is a symmetric negative definite matrix given that C_2 is nonsingular. The output error injection term $\nu(y, \hat{y}, \hat{z}, u, t)$ is defined by

$$\nu := k(y, \hat{z}, u, t) C_2^{-1} \frac{y - \hat{y}}{\|y - \hat{y}\|}, \quad \text{if } y - \hat{y} \neq 0, \tag{3.89}$$

where $k(y, \hat{z}, u, t)$ is a positive scalar function to be determined later.

Let $e_1 = z_1 - \hat{z}_1$ and $e_y = y - \hat{y} = C_2(z_2 - \hat{z}_2)$. Then, the error dynamical system is described by

$$\dot{e}_1 = F_1 e_1 + \begin{bmatrix} I_{n-p} & L \end{bmatrix} \left(G(T^{-1}z, u) - G(T^{-1}\hat{z}, u) \right), \tag{3.90}$$

$$
\begin{aligned}
\dot{e}_y &= C_2 A_3 e_1 + F e_y + C_2 \left(G_2(T^{-1}z, u) - G_2(T^{-1}\hat{z}, u) \right) \\
&\quad + C_2 E_2 \Psi(t, T^{-1}z, u) + C_2 D_2 f(y, u, t) - C_2 \nu.
\end{aligned}
\tag{3.91}
$$

Remark 3.2. The gain matrix K is introduced to guarantee that the following matrix $\begin{bmatrix} F_1 & 0 \\ C_2 A_3 & F \end{bmatrix}$ is stable. It can be directly obtained since the matrices F_1 and F in Eqs. (3.86) and (3.88) are both Hurwitz.

Proposition 3.3 [36]. *Suppose that Assumptions (3.1), (3.2), and (3.3) hold and the following matrix inequality*

$$\bar{A}^T \bar{P}^T + \bar{P}\bar{A} + \frac{1}{\varepsilon}\bar{P}\bar{P}^T + \varepsilon\gamma_G^2 I_{n-p} + \varrho P < 0 \tag{3.92}$$

is solvable for $\bar{P}$ where

$$\bar{P} := P\begin{bmatrix} I_{n-p} & L \end{bmatrix}, \quad \bar{A} := \begin{bmatrix} A_1 & A_3 \end{bmatrix}^T \tag{3.93}$$

for $P > 0$, ε and ϱ are some positive constants. γ_G is the Lipschitz constant for $G(x, u)$ with respect to x. Then, the error dynamical system (3.90) is asymptotically stable, i.e.

$$\|e_1(t)\| \leq M \|e_1(0)\| e^{-\frac{\varrho}{2}t}, \tag{3.94}$$

where $e_1(0)$ is the initial error and $M := \sqrt{\frac{\lambda_{\max(P)}}{\lambda_{\min(P)}}}$.

Remark 3.3. The inequality (3.92) can be transformed into a standard LMI feasibility problem: for a given scalar $\varrho > 0$, find matrices P, Y and a scalar ε such that:

$$\begin{bmatrix} PA_1 + A_1^T P + YA_3 + A_3^T Y^T + \varrho P + \varepsilon\gamma_G^2 & P & Y \\ P & -\varepsilon I_{n-p} & 0 \\ Y^T & 0 & -\varepsilon I_p \end{bmatrix} < 0, \tag{3.95}$$

where $Y := PL$ with $P > 0$. Moreover, a convex eigenvalue optimization problem can be posed which is to maximize γ_G by determining the values of P, Y and ε [36].

Proposition 3.4 [36]. *Suppose that Assumptions (3.1), (3.2), and (3.3) hold and the gain $k(\cdot)$ is chosen to satisfy*

$$k(y, \hat{z}, u, t) \geq (\|C_2 A_3\| + \|C_2\| \gamma_G)\hat{\omega}(t) + \|C_2 E_2\| \beta \\ + \|C_2 D_2\| \alpha(y, u, t) + \eta, \tag{3.96}$$

where η is a positive constant and $\hat{\omega}(t)$ has the following dynamics

$$\dot{\hat{\omega}}(t) = -\frac{1}{2}\varrho\hat{\omega}(t), \quad \hat{\omega}(0) \geq M \|e_1(0)\|. \tag{3.97}$$

Then, the system (3.91) is driven to the sliding surface

$$S = \{(e_1, e_y)|e_y = 0\},\qquad(3.98)$$

in finite time and remains on it thereafter.

It follows from Propositions 3.3 and 3.4 that system (3.87) is an SMO of system (3.85), where $\hat{y}$ is the observer output which will be used in the FDI. The proposed SMO will be analyzed in order to reconstruct or estimate the fault signal $f(y, u, t)$ in the presence of the uncertainty $\Psi(t, x, u)$.

Fault Reconstruction Via First-Order SMO

During the sliding motion, $e_y = \dot{e}_y = 0$. Since C_2 is nonsingular, it follows from Eq. (3.91) that

$$\begin{aligned} A_3 e_1 + \left(G_2(T^{-1}z, u) - G_2(T^{-1}\hat{z}, u)\right) \\ + E_2\Psi(t, T^{-1}z, u) + D_2 f(y, u, t) - \nu_{eq} &= 0, \end{aligned}\qquad(3.99)$$

where ν_{eq} is the equivalent output error injection signal, obtained from a low pass filter.

Since $\lim_{t\to\infty} e_1(t) = 0$ and

$$\lim_{t\to\infty} \left\| G_2(T^{-1}z, u) - G_2(T^{-1}\hat{z}, u) \right\| \le \gamma_G \left\| e_1(t) \right\| \to 0.$$

Thus, Eq. (3.99) can be written as

$$D_2 f(y, u, t) = \nu_{eq} - E_2\Psi(t, T^{-1}z, u) + d_1(t),\qquad(3.100)$$

where $\lim_{t\to\infty} d_1(t) = 0$. In the case when $\Psi(t, T^{-1}z, u) = 0$, the estimate of fault signal is

$$\hat{f}(t) = D_2^+ \nu_{eq},\qquad(3.101)$$

where D_2^+ is the pseudo-inverse of D_2, i.e. $D_2^+ D_2 = I_q$ and

$$\lim_{t\to\infty} \left(\hat{f}(t) - f(y, u, t)\right) \to 0.$$

In the case when $\Psi(t, T^{-1}z, u) \neq 0$, multiply both sides of Eq. (3.100) by D_2^+, it follows

$$f(t) = D_2^+ \nu_{eq} - D_2^+ E_2\Psi(t, T^{-1}z, u) + D_2^+ d_1(t).\qquad(3.102)$$

In view of Assumption 3.4 and Eq. (3.102), it follows

$$\left\| \hat{f}(t) - f(y, u, t) \right\| \leq \left\| D_2^+ E_2 \right\| \beta + \left\| D_2^+ d_1(t) \right\|, \tag{3.103}$$

where $\lim_{t \to \infty} D_2^+ d_1(t) = 0$.

The objective here is to choose an appropriate matrix D_2^+ such that the effect of the uncertainty $\Psi(t, T^{-1}z, u)$ is minimized, i.e. min $\left\| D_2^+ E_2 \right\|$. Let $\mathcal{D} = \{ X \in \mathbb{R}^{q \times p} \mid X D_2 = I_q \}$. The set $\mathcal{D}$ can be parameterized as

$$\mathcal{D} = \left\{ \left(D_2^T D_2 \right)^{-1} D_2^T + \mu D_2^N \mid \mu \in \mathbb{R}^{q \times (p-q)} \right\}, \tag{3.104}$$

where $D_2^N \in \mathbb{R}^{(p-q) \times p}$ spans the null-space of D_2 which implies that $D_2^N D_2 = 0$. Thus, for any $D_2^+ \in \mathcal{D}$, it follows

$$D_2^+ E_2 = \left(D_2^T D_2 \right)^{-1} D_2^T E_2 + \mu D_2^N E_2. \tag{3.105}$$

The objective is transformed into the following optimization problem

$$\min_{\mu \in \mathbb{R}^{q \times (p-q)}} = \left\{ \left\| \left(D_2^T D_2 \right)^{-1} D_2^T E_2 + \mu D_2^N E_2 \right\| \right\}. \tag{3.106}$$

This can be easily solved using LMI optimization approach [40].

Remark 3.4. In the case when $\Psi(t, T^{-1}z, u) = 0$, detection is inherent since precise reconstruction is achieved. However, when precise reconstruction is unavailable, detection is more difficult since the presence of uncertainty will make the equivalent output error injection be nonzero. Thus, it is difficult to distinguish the fault from the uncertainty. Provided the size of the bound $\left\| D_2^+ E_2 \right\| \beta$ is relatively small compared to the size of the fault, a reasonable solution is to set appropriate thresholds, and a level of detection can still be achieved.

3.4.2 SOSM Observer-Based FDI

In the above section, a standard first-order sliding mode algorithm was employed to force e_y and $\dot{e}_y$ to zero exactly in finite time. Then, the fault reconstruction is achieved by analyzing the output error injection which was generated by low pass filters. Therefore, it is an *approximate* method [6, 32]. However, there are two main drawbacks in the above method. The first one is that the approximation of the equivalent injections by low pass filters will typically introduce some delays, that lead to inaccurate estimates or to instability for high-order systems. The second one is that *a priori* information of the bound of the fault signal is not always available in practice. For the first problem, the solution is to replace the discontinuous injection signal by a continuous one. The solution of the second problem is to design an adaptive law for the gains of the SOSM algorithm. The gains of the algorithm stop increasing when the observation error converges to zero exactly.

In this section, state estimation, parameter identification and fault reconstruction are studied for a class of nonlinear systems with uncertain parameters, simultaneously. Adaptive-gain SOSM algorithms are employed handle the uncertainty with the *unknown* boundary by dynamically adapting their parameters. The approach involves a simple adaptive update law and the proposed adaptive-gain SOSM observer. The adaptive law is derived via the so-called "time scaling" approach [12], which are adapted dynamically according to the observation error. The uncertain parameters are estimated and then injected into an adaptive-gain SOSM observer, which maintains a sliding motion even in the presence of fault signals. Finally, once the sliding motion is achieved, the equivalent output error injection can be obtained directly and the fault signals are reconstructed based on this information.

3.4.2.1 Adaptive-Gain SOSML Design for FDI

Consider the following nonlinear system

$$
\begin{aligned}
\dot{x} &= Ax + g(x, u) + \phi(y, u)\theta + \omega(y, u)f(t), \\
y &= Cx,
\end{aligned}
\tag{3.107}
$$

where the matrix $A = \begin{bmatrix} A_1 & A_2 \\ A_3 & A_4 \end{bmatrix}$, $x \in \mathbb{R}^n$ is the system state vector, $u(t) \in \mathcal{U} \subset \mathbb{R}^m$ is the control input which is assumed to be known, $y \in \mathcal{Y} \subset \mathbb{R}^p$ is the output vector. The function $g(x, u) \in \mathbb{R}^n$ is Lipschitz continuous, $\phi(y, u) \in \mathbb{R}^{n \times q}$ and $\omega(y, u) \in \mathbb{R}^{n \times r}$ are assumed to be some smooth and bounded functions with $p \geq q + r$. The unknown parameter vector $\theta \in \mathbb{R}^q$ is assumed to be constant and $f(t) \in \mathbb{R}^r$ is a smooth fault signal vector, which satisfies

$$
\|f(t)\| \leq \rho_1, \quad \|\dot{f}(t)\| \leq \rho_2,
\tag{3.108}
$$

where ρ_1, ρ_2 are some positive constants that might be *known* or *unknown*.

Assume that (A, C) is an observable pair, and there exists a linear coordinate transformation $z = Tx = \begin{bmatrix} I_p & 0 \\ -H_{(n-p) \times p} & I_{n-p} \end{bmatrix} x = \begin{bmatrix} z_1^{\mathrm{T}} & z_2^{\mathrm{T}} \end{bmatrix}^{\mathrm{T}}$, with $z_1 \in \mathbb{R}^p$ and $z_2 \in \mathbb{R}^{n-p}$, such that

- $TAT^{-1} = \begin{bmatrix} A_{11} & A_{12} \\ A_{21} & A_{22} \end{bmatrix}$, where the matrix $A_{22} = A_4 - HA_2 \in \mathbb{R}^{(n-p) \times (n-p)}$ is Hurwitz stable.
- $CT^{-1} = \begin{bmatrix} I_p & 0 \end{bmatrix}$, where $I_p \in \mathbb{R}^{p \times p}$ is an identity matrix.

Assumption 3.5. There exists a function $\omega_1(y, u)$ such that

$$T\omega(y, u) = \begin{bmatrix} \omega_1(y, u) \\ 0 \end{bmatrix}, \tag{3.109}$$

where $\omega_1(y, u) \in \mathbb{R}^{p \times r}$.

Remark 3.5. Assumption 3.5 is a structural constraint on the fault distribution $\omega(\cdot, \cdot)$. It means that the faults only affect on the system output channel. It should be noted that there are no such structural constraints on the uncertain parameters distribution $\phi(\cdot, \cdot)$.

System (3.107) is described by the following equations in the new coordinate system,

$$\dot{z} = TAT^{-1}z + Tg(T^{-1}z, u) + T\phi(y, u)\theta + T\omega(y, u)f(t),$$
$$y = CT^{-1}z. \tag{3.110}$$

By reordering the state variables, the system (3.110) can be rewritten as

$$\dot{y} = A_{11}y + A_{21}z_2 + g_1(z_2, y, u) + \phi_1(y, u)\theta + \omega_1(y, u)f(t),$$
$$\dot{z}_2 = A_{22}z_2 + A_{21}y + g_2(z_2, y, u) + \phi_2(y, u)\theta,$$
$$y = z_1, \tag{3.111}$$

where

$$T\phi(y, u) = \begin{bmatrix} \phi_1(y, u) \\ \phi_2(y, u) \end{bmatrix}, \quad Tg(T^{-1}z, u) = \begin{bmatrix} g_1(z_2, y, u) \\ g_2(z_2, y, u) \end{bmatrix}, \tag{3.112}$$

$\phi_1(\cdot, \cdot) : \mathbb{R}^p \times \mathbb{R}^m \to \mathbb{R}^p$, $\phi_2(\cdot, \cdot) : \mathbb{R}^p \times \mathbb{R}^m \to \mathbb{R}^{n-p}$, $g_1(\cdot, \cdot, \cdot) : \mathbb{R}^p \times \mathbb{R}^{n-p} \times \mathbb{R}^m \to \mathbb{R}^p$ and $g_2(\cdot, \cdot, \cdot) : \mathbb{R}^p \times \mathbb{R}^{n-p} \times \mathbb{R}^m \to \mathbb{R}^{n-p}$.

We now consider the problem of an adaptive SOSM observer for system (3.111), in which the uncertain parameter is estimated with the help of an adaptive law. Then, an SOSM observer with gain adaptation is developed using the estimated parameter. Finally, based on the adaptive SOSM observer, a fault reconstruction method which can be implemented online is proposed. The basic assumption on the system (3.111) is as follows:

Assumption 3.6. There exists a nonsingular matrix $\bar{T} \in \mathbb{R}^{p \times p}$, such that

$$\bar{T} \begin{bmatrix} \phi_1(y, u) & \omega_1(y, u) \end{bmatrix} = \begin{bmatrix} \Phi_1(y, u) & 0_{q \times r} \\ 0_{r \times q} & \Phi_2(y, u) \end{bmatrix}, \tag{3.113}$$

where $\Phi_1(y, u) \in \mathbb{R}^{q \times q}$, $\Phi_2(y, u) \in \mathbb{R}^{r \times r}$ are both nonsingular matrices and bounded in $(y, u) \in \mathcal{Y} \times \mathcal{U}$.

Remark 3.6. The main limitation in Assumption 3.6 is that the matrix $\left[\phi_1(y, u),\right.$ $\left.\omega_1(y, u)\right]$ must be block-diagonalizable by elementary row transformations [37]. For the sake of simplicity, the case of only one fault signal and one uncertain parameter is considered ($q = r = 1$).

Let $z_y = \bar{T} y$, where $\bar{T}$ is defined in Assumption 3.6. Then, the system in (3.111) can be described by

$$
\begin{aligned}
\dot{z}_y &= \bar{T} A_{11} y + \bar{T} A_{21} z_2 + \bar{T} g_1(y, z_2, u) + \begin{bmatrix} \Phi_1(y, u) \\ 0 \end{bmatrix} \theta + \begin{bmatrix} 0 \\ \Phi_2(y, u) \end{bmatrix} f(t), \\
\dot{z}_2 &= A_{22} z_2 + A_{21} y + g_2(z_2, y, u) + \phi_2(y, u)\theta, \\
y &= \bar{T}^{-1} z_y,
\end{aligned}
\tag{3.114}
$$

where

$$
\bar{T} \cdot A_{11} = \begin{bmatrix} \bar{A}_{11} \\ \bar{A}_{12} \end{bmatrix}, \quad \bar{T} \cdot A_{12} = \begin{bmatrix} \bar{A}_{21} \\ \bar{A}_{22} \end{bmatrix}, \quad \bar{T} \cdot g_1(y, z_2, u) = \begin{bmatrix} W_{g_1}(y, z_2, u) \\ W_{g_2}(y, z_2, u) \end{bmatrix}.
$$

Let us define $z_y = \left[z_{y_1}, z_{y_2}\right]^{\mathsf{T}}$, where $z_{y_1} \in \mathbb{R}^q$, $z_{y_2} \in \mathbb{R}^r$. Then, in view of (3.114), we can obtain

$$
\begin{aligned}
\dot{z}_{y_1} &= \bar{A}_{11} y + \bar{A}_{21} z_2 + W_{g_1}(y, z_2, u) + \Phi_1(y, u)\theta, \\
\dot{z}_{y_2} &= \bar{A}_{12} y + \bar{A}_{22} z_2 + W_{g_2}(y, z_2, u) + \Phi_2(y, u) f(t), \\
\dot{z}_2 &= A_{21} y + A_{22} z_2 + g_2(y, z_2, u) + \phi_2(y, u)\theta, \\
y &= T^{-1} \left[z_{y_1}\ z_{y_2}\right]^{\mathsf{T}}.
\end{aligned}
\tag{3.115}
$$

Regarding the above system (3.115), the adaptive SOSM observer is represented by the following dynamical system

$$
\begin{aligned}
\dot{\hat{z}}_{y_1} &= \bar{A}_{11} y + \bar{A}_{21} \hat{z}_2 + W_{g_1}(y, \hat{z}_2, u) + \Phi_1(y, u)\hat{\theta} + \mu(e_{y_1}), \\
\dot{\hat{z}}_{y_2} &= \bar{A}_{12} y + \bar{A}_{22} \hat{z}_2 + W_{g_2}(y, \hat{z}_2, u) + \mu(e_{y_2}), \\
\dot{\hat{z}}_2 &= A_{21} y + A_{22} \hat{z}_2 + g_2(y, \hat{z}_2, u) + \phi_2(y, u)\hat{\theta},
\end{aligned}
\tag{3.116}
$$

where $\mu(\cdot)$ is the SOSM algorithm

$$
\mu(s) = \lambda(t)|s|^{\frac{1}{2}} \text{sign}(s) + \alpha(t) \int_0^t \text{sign}(s) d\tau + k_\lambda(t)s + k_\alpha(t) \int_0^t s d\tau,
$$

and the adaptive gains $\lambda(t)$, $\alpha(t)$, $k_\lambda(t)$ and $k_\alpha(t)$ are determined later.

The observation errors are defined as $e_{y_1} = z_{y_1} - \hat{z}_{y_1}$, $e_{y_2} = z_{y_2} - \hat{z}_{y_2}$, $e_2 = z_2 - \hat{z}_2$, $\tilde{\theta} = \theta - \hat{\theta}$. The estimate of θ, denoted by $\hat{\theta}$, is given by the following adaptive law

$$\dot{\hat{\theta}} = -K(y, u)\left(\bar{A}_{11}y + \bar{A}_{21}\hat{z}_2 + W_{g_1}(y, \hat{z}_2, u) + \Phi_1(y, u)\hat{\theta} - \dot{z}_{y_1}\right), \quad (3.117)$$

where $K(y, u)$ is a positive design matrix which will be determined later.

Remark 3.7. It can be seen that the adaptive law (3.117) depends upon $\dot{z}_{y_1}$. A real-time robust exact differentiator proposed in [17] can be used to estimate the time derivative of z_{y_1} in finite time. The differentiator has the following form

$$\dot{z}_0 = -\lambda_0 L_0^{\frac{1}{2}}|z_0 - z_{y_1}|^{\frac{1}{2}}\text{sign}(z_0 - z_{y_1}) + z_1,$$
$$\dot{z}_1 = -\alpha_0 L_0 \text{sign}(z_0 - z_{y_1}), \quad (3.118)$$

where z_0 and z_1 are the real-time estimations of z_{y_1} and $\dot{z}_{y_1}$ respectively. The parameters of the differentiator $\lambda_0 = 1$, $\alpha_0 = 1.1$ are suggested in [17]. L_0 is the only parameter needs to be tuned according to the condition $|\ddot{z}_{y_1}| \le L_0$.

Subtracting (3.116) from (3.115), the error dynamical equation is described by

$$\dot{e}_2 = A_{22}e_2 + \tilde{g}_2(y, z_2, \hat{z}_2, u) + \phi_2(y, u)\tilde{\theta}, \quad (3.119)$$
$$\dot{\tilde{\theta}} = -K(y, u)\left(\bar{A}_{21}e_2 + \tilde{W}_{g_1}(y, z_2, \hat{z}_2, u) + \Phi_1(y, u)\tilde{\theta}\right), \quad (3.120)$$
$$\dot{e}_{y_1} = -\mu(e_{y_1}) + \bar{A}_{21}e_2 + \Phi_1(y, u)\tilde{\theta} + \tilde{W}_{g_1}(y, z_2, \hat{z}_2, u), \quad (3.121)$$
$$\dot{e}_{y_2} = -\mu(e_{y_2}) + \bar{A}_{22}e_2 + \Phi_2(y, u)f(t) + \tilde{W}_{g_2}(y, z_2, \hat{z}_2, u), \quad (3.122)$$

where

$$\tilde{g}_2(y, z_2, \hat{z}_2, u) = g_2(y, z_2, u) - g_2(y, \hat{z}_2, u)$$
$$\tilde{W}_{g_1}(y, z_2, \hat{z}_2, u) = W_{g_1}(y, z_2, u) - W_{g_1}(y, \hat{z}_2, u)$$
$$\tilde{W}_{g_2}(y, z_2, \hat{z}_2, u) = W_{g_2}(y, z_2, u) - W_{g_2}(y, \hat{z}_2, u)$$

Some assumptions are imposed upon the error dynamical systems (3.119)–(3.122).

Assumption 3.7. The known nonlinear terms $g_2(y, z_2, u)$, $W_{g_1}(y, z_2, u)$ and $W_{g_2}(y, z_2, u)$ are Lipschitz continuous with respect to z_2

$$\|g_2(y, z_2, u) - g_2(y, \hat{z}_2, u)\| \le \gamma_2\|z_2 - \hat{z}_2\|, \quad (3.123)$$
$$\|W_{g_1}(y, z_2, u) - W_{g_1}(y, \hat{z}_2, u)\| \le \gamma_{g_1}\|z_2 - \hat{z}_2\|, \quad (3.124)$$
$$\|W_{g_2}(y, z_2, u) - W_{g_2}(y, \hat{z}_2, u)\| \le \gamma_{g_2}\|z_2 - \hat{z}_2\|, \quad (3.125)$$

where $\gamma_{g_1}, \gamma_{g_2}$ and γ_2 are the known Lipschitz constants of $W_{g_1}(y, z_2, u), W_{g_2}(y, z_2, u)$ and $g_2(y, z_2, u)$, respectively [39].

Assumption 3.8. Assume that the Hurwitz matrix A_{22} satisfies the following Riccati equation

$$A_{22}^{\mathrm{T}}P_1 + P_1A_{22} + \gamma_2^2 P_1P_1 + (2+\varepsilon)I_{n-p} = 0, \qquad (3.126)$$

which has a SPD solution P_1 for some $\varepsilon > 0$ [24].

Assumption 3.9. Assume that the positive design matrix $K(y,u)$ satisfies the following equation

$$K(y,u)\Phi_1(y,u) + \Phi_1^{\mathrm{T}}(y,u)K^{\mathrm{T}}(y,u) - \gamma_{g_1}^2 K(y,u)K^{\mathrm{T}}(y,u) - \epsilon I_q = 0, \quad (3.127)$$

for some $\epsilon > 0$.

Assumption 3.10. It is assumed that $\|\Phi_1(y,u)\|$, $\|\Phi_2(y,u)\|$ are bounded in $(y,u) \in \mathcal{Y} \times \mathcal{U}$.

Now, we will first consider the stability of the error systems (3.119) and (3.120).

Theorem 3.2. *Consider systems (3.119) and (3.120) satisfying Assumptions (3.7)–(3.10). Then, the error systems (3.119, 3.120) are exponentially stable, if for any $(y,u) \in \mathcal{Y} \times \mathcal{U}$, the following matrix*

$$Q_1 = \begin{bmatrix} \varepsilon I_{n-p} & P_1\phi_2(y,u) - \bar{A}_{21}^{\mathrm{T}}K^{\mathrm{T}}(y,u) \\ \phi_2^{\mathrm{T}}(y,u)P_1 - K(y,u)\bar{A}_{21} & \epsilon I_q \end{bmatrix}, \quad (3.128)$$

is positive definite.

Proof. A candidate Lyapunov function is chosen as

$$V_1(e_2, \tilde{\theta}) = e_2^{\mathrm{T}}P_1e_2 + \tilde{\theta}^{\mathrm{T}}\tilde{\theta}, \qquad (3.129)$$

and the time derivative of V_1 along the solution of the system (3.119) and (3.120) is given by

$$\begin{aligned}
\dot{V}_1 &= e_2^{\mathrm{T}}(A_{22}^{\mathrm{T}}P_1 + P_1A_{22})e_2 - 2e_2^{\mathrm{T}}\bar{A}_{21}^{\mathrm{T}}K^{\mathrm{T}}(y,u)\tilde{\theta} + 2e_2^{\mathrm{T}}P_1\phi_2(y,u)\tilde{\theta} \\
&\quad + 2e_2^{\mathrm{T}}P_1\tilde{g}_2(y,z_2,\hat{z}_2,u) - 2\tilde{\theta}^{\mathrm{T}}K(y,u)\tilde{W}_{g_1}(y,z_2,\hat{z}_2,u) \\
&\quad - \tilde{\theta}^{\mathrm{T}}\left(K(y,u)\Phi_1(y,u) + \Phi_1^{\mathrm{T}}(y,u)K^{\mathrm{T}}(y,u)\right)\tilde{\theta} \\
&\leq e_2^{\mathrm{T}}(A_{22}^{\mathrm{T}}P_1 + P_1A_{22} + \gamma_2^2 P_1P_1 + 2I_{n-p})e_2 \\
&\quad + 2e_2^{\mathrm{T}}\left(P_1\phi_2(y,u) - \bar{A}_{21}^{\mathrm{T}}K^{\mathrm{T}}(y,u)\right)\tilde{\theta} - \epsilon\tilde{\theta}^{\mathrm{T}}\tilde{\theta} \\
&= -\left[e_2^{\mathrm{T}}\ \tilde{\theta}^{\mathrm{T}}\right]Q_1\begin{bmatrix} e_2 \\ \tilde{\theta} \end{bmatrix}.
\end{aligned}$$

Hence, the conclusion follows from the assumption that Q_1 given in (3.128) is positive definitive in $(y,u) \in \mathcal{Y} \times \mathcal{U}$.

Remark 3.8. Theorem 3.2 shows that $\lim\limits_{t \to \infty} e_2(t) = 0$ and $\lim\limits_{t \to \infty} \tilde{\theta}(t) = 0$. Conse-quently, the errors $e_2, \tilde{\theta}$ and its derivatives $\dot{e}_2, \dot{\tilde{\theta}}$ are bounded. Under Assumptions (3.7) and (3.10), the time derivatives of the nonlinear terms in the error dynamics (3.121) and (3.122) are bounded:

$$\left\| \frac{d}{dt} \left(\bar{A}_{22} e_2 + \Phi_2(y, u) f(t) + \tilde{W}_{g_2}(y, z_2, \hat{z}_2, u) \right) \right\| \leq \chi_1,$$
$$\left\| \frac{d}{dt} \left(\bar{A}_{21} e_2 + \Phi_1(y, u) \tilde{\theta} + \tilde{W}_{g_1}(y, z_2, \hat{z}_2, u) \right) \right\| \leq \chi_2, \tag{3.130}$$

where χ_1 and χ_2 are some *unknown* positive constants.

According to the results of Theorem 3.1, the finite time convergence of the systems (3.121) and (3.122) can be obtained directly. Theorems (3.1) and (3.2) have shown that systems (3.116) and (3.117) are an asymptotic state observer and an uncertain parameter observer for the system (3.115), respectively. In the next part, we will develop the fault reconstruction approach based on those two observers.

3.4.2.2   Fault Reconstruction via SOSM Observer

The fault signal $f(t)$ will be reconstructed based on the proposed observer by using an equivalent output error injection which can be obtained once the sliding surface is reached and maintained on it thereafter.

It follows from Theorem 3.1 that e_{y_2} and $\dot{e}_{y_2}$ in (3.122) are driven to zero in finite time. Thus, the equivalent output error injection can be obtained directly

$$\mu(e_{y_2}) = \bar{A}_{22} e_2 + \Phi_2(y, u) f(t) + \tilde{W}_{g_2}(y, z_2, \hat{z}_2, u). \tag{3.131}$$

From Assumption 3.6, $\Phi_2(y, u)$ is a bounded nonsingular matrix in $(y, u) \in \mathcal{Y} \times \mathcal{U}$ and $\lim_{t \to \infty} \tilde{W}_{g_2}(y, z_2, \hat{z}_2, u) = 0$, then the estimate of $f(t)$ can be constructed as

$$\hat{f}(t) = \Phi_2^{-1}(y, u) \mu(e_{y_2}). \tag{3.132}$$

Theorem 3.3. *Suppose that conditions of Theorems (3.1) and (3.2) are satisfied, then $\hat{f}(t)$ defined in (3.132) is a precise reconstruction of the fault signal $f(t)$ since*

$$\lim_{t \to \infty} \| f(t) - \hat{f}(t) \| = 0. \tag{3.133}$$

Proof. It follows from (3.131) and (3.132) that

$$\| f(t) - \hat{f}(t) \| = \| \Phi_2^{-1}(y, u)(\bar{A}_{22} e_2 + \tilde{W}_{g_2}) \|$$
$$\leq \| \Phi_2^{-1}(y, u) \bar{A}_{22} \| \| e_2 \| + \gamma_{g_2} \| \Phi_2^{-1}(y, u) \| \| e_2 \|. \tag{3.134}$$

From Theorem 3.2, $\lim\limits_{t\to\infty} \|e_2\| = 0$, it follows that

$$\lim_{t\to\infty} \|f(t) - \hat{f}(t)\| = 0. \tag{3.135}$$

Hence, Theorem 3.3 is proven. ∎

3.5 Illustrative Examples

Let us now see some applicative examples of SOSM observer design and its application to FDI.

In this example, we shall briefly outline the application of observers described in this chapter to a class of nonlinear systems. Consider the following nonlinear system [18],

$$\begin{aligned}
\dot{x}_1 &= -x_1 - x_1^2 - x_2 + x_3 + \omega(t), \\
\dot{x}_2 &= -x_1 - 3x_1^2 + 2x_1^3 + 2x_1x_2 + 2x_1x_3 - 2x_1\omega(t), \\
\dot{x}_3 &= x_1 + x_1^2 - x_2 - x_3 - x_3^2, \\
y^{\mathrm{T}} &= \begin{bmatrix} y_1 & y_2 \end{bmatrix} = \begin{bmatrix} x_1 & x_3 \end{bmatrix},
\end{aligned} \tag{3.136}$$

where $x^{\mathrm{T}} = \begin{bmatrix} x_1 & x_2 & x_3 \end{bmatrix}$ are the states, $\omega(t) = 2 \cdot \sin(5t)$ is an unknown disturbance, and y are the measurable outputs. Define the following coordinate transformation,

$$z = \Phi(x) = \begin{bmatrix} x_1 \\ x_1^2 + x_2 \\ x_3 \end{bmatrix}. \tag{3.137}$$

The system (3.136) can be transformed into the following system

$$\begin{aligned}
\dot{z}_1 &= -z_1 - z_2 + z_3 + \omega(t), \\
\dot{z}_2 &= -z_1 + 4z_1z_3 - 5z_1^2, \\
\dot{z}_3 &= z_1 - z_2 - z_3 + 2z_1^2 - z_3^2, \\
y^{\mathrm{T}} &= \begin{bmatrix} z_1 & z_3 \end{bmatrix}.
\end{aligned} \tag{3.138}$$

The parameters for the simulation are shown in Table 3.1.

Table 3.1 Parameters for the numerical simulation

Original states	Values	Estimate states	Values
$x_1(0)$	0	$\hat{x}_1(0)$	0.3
$x_2(0)$	0	$\hat{x}_2(0)$	0.39
$x_3(0)$	0	$\hat{x}_3(0)$	0.3

3.5.1 Adaptive-Gain SOSM Observer Design

The adaptive-gain SOSM observer is designed as follows

$$
\begin{aligned}
\dot{\hat{z}}_1 &= -y_1 + y_2 + \nu(e_{y_1}), \\
\dot{\hat{z}}_2 &= -y_1 + 4y_1 y_2 - 5y_1^2 + E_1 \nu(\tilde{z}_2 - \hat{z}_2), \\
\dot{\hat{z}}_3 &= y_1 + 2y_1^2 - y_2 - y_2^2 + \nu(e_{y_2}),
\end{aligned}
\tag{3.139}
$$

where $e_{y_1} = y_1 - \hat{z}_1$, $e_{y_2} = y_2 - \hat{z}_3$, the error injection term $\nu(\cdot)$ is obtained from the adaptive gain SOSM algorithm (3.60), (3.62), and (3.63) and

$$
\tilde{z}_2 = -\nu(e_{y_2}), \quad E_1 = \begin{cases} 1, & \text{if } |e_{y_2}| \le \epsilon, \\ 0, & \text{otherwise,} \end{cases}
\tag{3.140}
$$

where ϵ is a small positive constant.

Denote $e_2 = z_2 - \hat{z}_2$, the error dynamical system is given as follows:

$$
\begin{aligned}
\dot{e}_{y_1} &= -\nu(e_{y_1}) - z_2 + w(t), \\
\dot{e}_{y_2} &= -\nu(e_{y_2}) - z_2, \\
\dot{e}_2 &= -E_1 \nu_3(\tilde{z}_2 - \hat{z}_2).
\end{aligned}
\tag{3.141}
$$

With appropriate gains of the adaptive gain SOSM algorithm, the error dynamical system (3.141) converges to zero in finite time. Moreover, the estimate of the disturbance is obtained as $\hat{w}(t) = \nu(e_{y_1}) - \nu(e_{y_2})$. The state observation and disturbance estimation is shown in Fig. 3.3. It can be seen from Fig. 3.4 that the adaptive law (3.63) is effective under disturbance effect.

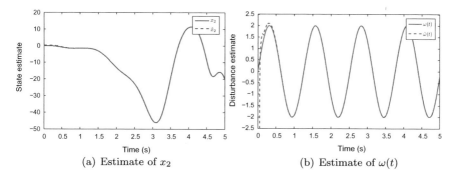

(a) Estimate of x_2 (b) Estimate of $\omega(t)$

Fig. 3.3 Estimates of state and disturbance

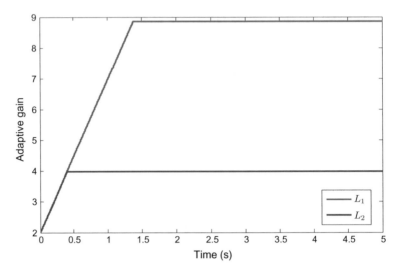

Fig. 3.4 Adaptive law $L(t)$ of the SOSM algorithm (3.63)

3.5.2 SOSM Observer Based FDI

An example is given in order to illustrate the design of adaptive-gain SOSM observer-based fault reconstruction scheme. Consider a pendulum system [5]

$$\dot{x}_1 = x_2 + f(t),$$
$$\dot{x}_2 = \frac{1}{J}u - \frac{g}{L}\sin(x_1) - \frac{V_s}{J}x_2,$$
$$y = x_1, \qquad\qquad\qquad\qquad (3.142)$$

where x_1, x_2 are the angle of oscillation and angular velocity, respectively. $M = 1.1$ kg is the pendulum mass, $g = 9.815$ m/s^2 is the gravitational force, $L = 1$ m is the pendulum length, $J = ML^2 = 1.1$ kg $\cdot$ m^2 is the arm inertia, $V_s = 1.8$ kg $\cdot$ m/s^2 is the pendulum viscous friction coefficient and $f(t)$ is considered as a bounded fault signal. For simulation purposes, it was taken as

$$f(t) = 0.5\sin(2t) + 0.5\cos(5t). \qquad\qquad (3.143)$$

The controller u is known which is used to drive x_1 to the desired value $x_1^* = \sin(t)$

$$u = -30\text{sign}(s) - 25\text{sign}(\dot{s}), \qquad\qquad (3.144)$$

where $s = x_1 - x_1^*$.

The observer based on STA or SOSM algorithm is designed as

$$\dot{\hat{x}}_1 = \hat{x}_2 + \mu(e_y),$$
$$\dot{\hat{x}}_2 = \frac{1}{J}u - \frac{g}{L}\sin(x_1) - \frac{V_s}{J}\hat{x}_2,$$
$$\hat{y} = \hat{x}_1, \qquad (3.145)$$

where $e_y = y - \hat{y}$. Let $e_2 = x_2 - \hat{x}_2$, the error dynamics are given as follows

$$\dot{e}_y = -\mu(e_y) + e_2 + f(t),$$
$$\dot{e}_2 = -\frac{V_s}{J}e_2. \qquad (3.146)$$

During the sliding motion $e_y = \dot{e}_y = 0$, a fault reconstruction signal $\hat{f}(t) = \mu(e_y)$ is introduced given that $\lim_{t\to\infty} e_2(t) = 0$.

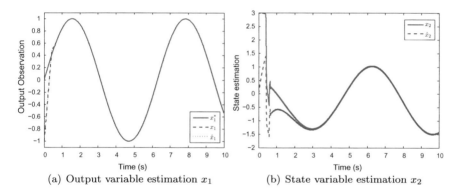

(a) Output variable estimation x_1 (b) State variable estimation x_2

Fig. 3.5 The performance of observer (3.145)

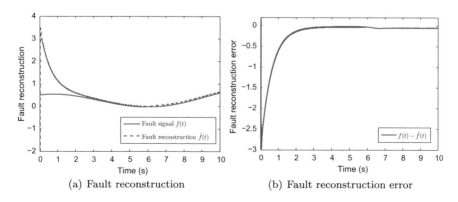

(a) Fault reconstruction   (b) Fault reconstruction error

Fig. 3.6 Fault reconstruction and its error for the pendulum system

The initial values $x_1(0) = -1$, $x_2(0) = 3$ are set for the pendulum and $\hat{x}_1(0) = \hat{x}_2(0) = 0$ are set for the two observers. The state estimations and fault reconstruction are shown in Figs. 3.5 and 3.6, respectively.

3.6 Conclusion

In this chapter, we have discussed the SMO design and its application to fault reconstruction. Traditional first order SMO design for linear uncertain systems has been introduced first and their drawbacks have been discussed. Then, SOSM algorithms have been developed to generate continuous output error injection signals, which eliminate the use of any low pass filters and their applications to fault reconstruction have also been presented. Finally, our contributions in adaptive-gain SOSM observer and its use in FDI have been introduced. An adaptive-gain SOSM algorithm has been proposed, which does not require the *a priori* knowledge of the uncertainty's boundary. Additionally, adaptive-gain SOSM observer based fault reconstruction has been considered for a class of nonlinear systems with uncertain parameters. State estimation, parameter identification and fault reconstruction are performed, simultaneously. Finally, two illustrative examples have been provided to show both the observer designs and their applications to fault reconstruction.

In the subsequent chapters, we shall concentrate on the applicative aspect of the above discussed sliding mode algorithms. Namely, a complex nonlinear system, i.e., PEMFC will be presented and the observation and FDI of its subsystems will be discussed.

References

1. Bacciotti, A., Rosier, L.: Liapunov Functions and Stability in Control Theory. Springer (2005)
2. Besançon, G.: High-gain observation with disturbance attenuation and application to robust fault detection. Automatica **39**(6), 1095–1102 (2003)
3. Chen, J., Patton, R., Zhang, H.: Design of unknown input observers and robust fault detection filters. Int. J. Control. **63**(1), 85–105 (1996)
4. Dávila, A., Moreno, J., Fridman, L.: Optimal Lyapunov function selection for reaching time estimation of super twisting algorithm. In: Proceedings of the 48th IEEE Conference on Decision and Control, held jointly with the 28th Chinese Control Conference (CDC/CCC), pp. 8405–8410. IEEE (2009)
5. Davila, J., Fridman, L., Levant, A., et al.: Second-order sliding-mode observer for mechanical systems. IEEE Trans. Autom. Control. **50**(11), 1785–1789 (2005)
6. Edwards, C., Spurgeon, S.K., Patton, R.J.: Sliding mode observers for fault detection and isolation. Automatica **36**(4), 541–553 (2000)
7. Edwards, C., Tan, C.P.: Sensor fault tolerant control using sliding mode observers. Control. Eng. Pract. **14**(8), 897–908 (2006)
8. Filippov, A., Arscott, F.: Differential Equations with Discontinuous Righthand Sides: Control Systems, vol. 18. Springer (1988)

9. Floquet, T., Barbot, J.P.: Super twisting algorithm-based step-by-step sliding mode observers for nonlinear systems with unknown inputs. Int. J. Syst. Sci. **38**(10), 803–815 (2007)
10. Floquet, T., Edwards, C., Spurgeon, S.K.: On sliding mode observers for systems with unknown inputs. Int. J. Adapt. Control. Signal Process. **21**(8–9), 638–656 (2007)
11. Gonzalez, T., Moreno, J.A., Fridman, L.: Variable gain super-twisting sliding mode control. IEEE Trans. Autom. Control. **57**(8), 2100–2105 (2012)
12. Hong, Y., Wang, J., Xi, Z.: Stabilization of uncertain chained form systems within finite settling time. IEEE Trans. Autom. Control. **50**(9), 1379–1384 (2005)
13. Hou, M., Muller, P.: Design of observers for linear systems with unknown inputs. IEEE Trans. Autom. Control. **37**(6), 871–875 (1992)
14. Jiang, B., Staroswiecki, M., Cocquempot, V.: Fault estimation in nonlinear uncertain systems using robust/sliding-mode observers. IEE Proc.-Control. Theory Appl. **151**(1), 29–37 (2004)
15. Khalil, H.K.: Nonlinear Systems. Prentice Hall (2001)
16. Kobayashi, S., Furuta, K.: Frequency characteristics of Levant's differentiator and adaptive sliding mode differentiator. Int. J. Syst. Sci. **38**(10), 825–832 (2007)
17. Levant, A.: Robust exact differentiation via sliding mode technique. Automatica **34**(3), 379–384 (1998)
18. Liu, J., Laghrouche, S., Wack, M.: Finite time observer design for a class of nonlinear systems with unknown inputs. In: American Control Conference (ACC), pp. 286–291. IEEE (2013)
19. Moreno, J., Osorio, M.: Strict Lyapunov functions for the super-twisting algorithm. IEEE Trans. Autom. Control. **57**(4), 1035–1040 (2012)
20. Moreno, J.A., Osorio, M.: A Lyapunov approach to second-order sliding mode controllers and observers. In: 47th IEEE Conference on Decision and Control(CDC), pp. 2856–2861. IEEE (2008)
21. Orlov, Y.: Finite time stability and robust control synthesis of uncertain switched systems. SIAM J. Control. Optim. **43**(4), 1253–1271 (2004)
22. Persis, C.D., Isidori, A.: A geometric approach to nonlinear fault detection and isolation. IEEE Trans. Autom. Control. **46**(6), 853–865 (2001)
23. Qiu, Z., Gertler, J.: Robust FDI systems and H_∞-optimization-disturbances and tall fault case. In: Proceedings of the 32nd IEEE Conference on Decision and Control, USA, pp. 1710–1715 (1993)
24. Rajamani, R.: Observers for Lipschitz nonlinear systems. IEEE Trans. Autom. Control. **43**(3), 397–401 (1998)
25. Respondek, W., Pogromsky, A., Nijmeijer, H.: Time scaling for observer design with linearizable error dynamics. Automatica **40**(2), 277–285 (2004)
26. Saif, M., Guan, Y.: A new approach to robust fault detection and identification. IEEE Trans. Aerosp. Electron. Syst. **29**(3), 685–695 (1993)
27. Slotine, J.J., Hedrick, J., Misawa, E.: On sliding observers for nonlinear systems. In: American Control Conference, pp. 1794–1800. IEEE (1986)
28. Slotine, J.J.E., Hedrick, J.K., Misawa, E.A.: On sliding observers for nonlinear systems. ASME J. Dyn. Syst. Meas. Control. **109**, 245–252 (1987)
29. Spurgeon, S.K.: Sliding mode observers: a survey. Int. J. Syst. Sci. **39**(8), 751–764 (2008)
30. Tan, C., Edwards, C.: Sliding mode observers for detection and reconstruction of sensor faults. Automatica **38**(10), 1815–1821 (2002)
31. Tan, C., Edwards, C.: Sliding mode observers for robust detection and reconstruction of actuator and sensor faults. Int. J. Robust Nonlinear Control. **13**(5), 443–463 (2003)
32. Utkin, V.I.: Sliding Modes in Control and Optimization. Springer, Berlin (1992)
33. Veluvolu, K.C., Defoort, M., Soh, Y.C.: High-gain observer with sliding mode for nonlinear state estimation and fault reconstruction. J. Frankl. Inst. **351**(4), 1995–2014 (2014)
34. Walcott, B., Żak, S.: State observation of nonlinear uncertain dynamical systems. IEEE Trans. Autom. Control. **32**(2), 166–170 (1987)
35. Wang, H., Huang, Z.J., Daley, S.: On the use of adaptive updating rules for actuator and sensor fault diagnosis. Automatica **33**, 217–225 (1997)

36. Yan, X., Edwards, C.: Nonlinear robust fault reconstruction and estimation using a sliding mode observer. Automatica **43**(9), 1605–1614 (2007)
37. Yan, X., Edwards, C.: Adaptive sliding-mode-observer-based fault reconstruction for nonlinear systems with parametric uncertainties. IEEE Trans. Ind. Electron. **55**(11), 4029–4036 (2008)
38. Yang, H., Saif, M.: Nonlinear adaptive observer design for fault detection. In: Proceedings of the American Control Conference, USA, pp. 1136–1139 (1995)
39. Zhang, X., Polycarpou, M., Parisini, T.: Fault diagnosis of a class of nonlinear uncertain systems with lipschitz nonlinearities using adaptive estimation. Automatica **46**(2), 290–299 (2010)
40. Zhou, K., Doyle, J.C., Glover, K., et al.: Robust and optimal control, vol. 40. Prentice Hall, New Jersey (1996)
41. Zubov, V.: Methods of A.M. Lyapunov and Their Application. P. Noordhoff (1964)

Chapter 4
Introduction of Proton Exchange Membrane Fuel Cell Systems

This chapter introduces some fuel cells classified into five different categories based on the electrolyte chemistry. The power generation mechanism of the PEMFC is presented. The PEMFC air-feed system model including supply manifold model, compressor model, cathode flow model, etc. is detailed. For an understandable SMC scenario design to the PEMFC, an experimental validation performed through the HIL test bench is provided, based on a commercial twin screw compressor and a real-time PEMFC emulator.

4.1 Introduction of PEMFCs

Fuel cells are electrochemical devices that convert the chemical energy of a gaseous fuel directly into electricity. They are under intensive development in the past few years as they are regarded as an efficient carbon free electricity production technology. The basic principle of the fuel cell was first discovered by Grove [3] in 1839, through an experiment that demonstrated that the reaction of hydrogen and oxygen produces electrical current. Modern fuel cells consist of an electrolyte sandwiched between two electrodes as shown in Fig. 4.1.

Fuel cells can be classified into five different categories based on the electrolyte chemistry: PEMFC, solid oxide fuel cell (SOFC), molten carbonate fuel cell (MCFC), phosphoric acid fuel cell (PAFC) and aqueous alkaline fuel cell (AFC). Among these kinds of fuel cells, PEMFCs are suitable for automobile applications due to high energy density, low working temperature, simple structure and long life as well as low corrosion [11]. Table 4.1 provides a summary of various fuel cell types and corresponding characteristics.

The fuel cell under consideration throughout this monograph is the PEMFC. The electrolyte membrane of a PEMFC has a special property that allows only positive protons to pass through while blocking electrons. Hydrogen molecules are split into

© Springer Nature Switzerland AG 2020
J. Liu et al., *Sliding Mode Control Methodology in the Applications of Industrial Power Systems*, Studies in Systems, Decision and Control 249,
https://doi.org/10.1007/978-3-030-30655-7_4

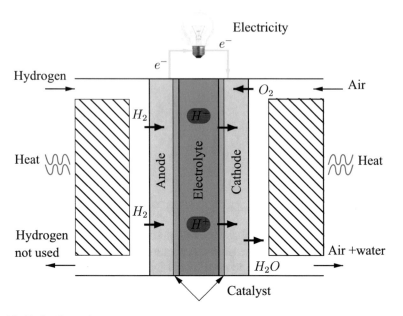

Fig. 4.1 Fuel cell reaction

Table 4.1 Summary of typical fuel cell characteristics (Table 16, [5])

Electrolyte material	Operating temperature	Anticipated applications	Comments
PEMFC	80 °C	Stationary and automobile	Minimum contamination and material problem
AFC	Approx 100 °C	Space program	Susceptible to contamination, very expensive
PAFC	Approx 100 °C	Stationary	Higher temperature and longer warm up time makes unsuitable for vehicles
SOFC	1000 °C	Stationary	Very high temperature create material problems, steam generation could increase efficiency by cogeneration
MCFC	600 °C	Stationary	Same as SOFC

protons and free electrons at the anode. These protons flow to the cathode through the electrolyte and react with the supplied oxygen and return electrons to produce water. During this process, the electrons pass through an external load circuit and provide electricity. There are two important subsystems required for proper operation of PEMFCs. First is the air-feed system that supplies oxygen to the cathode and therefore indirectly regulates the net power output of the fuel cell. Then, there is the power converter system that forms a link between the fuel cell output and the power bus of the power system. This chapter is dedicated to the description and modeling of the air-feed system. We will start by a short description of the power generation in the PEMFC and highlight the importance of air-feed system in it.

4.2 PEMFC Stack Voltage

In order to understand the importance of the air-feed system, let us look at the power generation mechanism of the PEMFC. A typical PEMFC polarization curve is shown in Fig. 4.2. The cell voltage is modeled from its static characteristic, which is a function of stack current, cathode pressure, reactant partial pressures, fuel cell temperature and membrane humidity [11]

$$v_{fc} = E - v_{act} - v_{ohm} - v_{conc}. \tag{4.1}$$

The open circuit voltage E can be calculated as

$$E = 1.229 - 0.85 \cdot 10^{-3} \left(T_{fc} - 298.15 \right)$$
$$+ 4.3085 \cdot 10^{-5} T_{fc} \left[\ln \left(p_{H_2} \right) + \frac{1}{2} \ln \left(p_{O_2} \right) \right], \tag{4.2}$$

where T_{fc} is the temperature of the FC (in Kelvin), p_{H_2} and p_{O_2} are the partial pressures of hydrogen and oxygen, respectively (in bar). It can be seen that

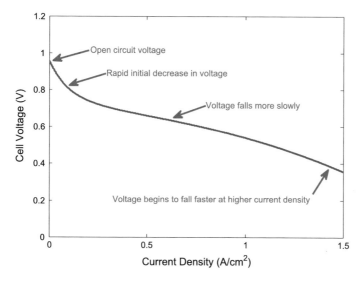

Fig. 4.2 Typical fuel cell voltage

the air-feed system indirectly controls the output voltage and hence the power, as it controls the oxygen pressure in the cathode. The current density i is defined as

$$i = \frac{I_{st}}{A_{fc}},$$ (4.3)

where I_{st}(A) is the stack current and A_{fc}(cm^2) is the active area. The activation loss v_{act}, ohmic loss v_{ohm} and concentration loss v_{conc} are expressed as follows:

1. $v_{act} = a \ln\left(\frac{i}{i_0}\right)$ is due to the difference between the velocity of the reactions in the anode and cathode [6], a and i_0 are constants which can be determined empirically. It should be noted that this equation is only valid for $i > i_0$. Therefore, a similar function that is valid for the entire range of i is preferred: $v_{act} = v_0 + v_a\left(1 - e^{-b_1 i}\right)$, v_0 (volts) is the voltage drop at zero current density, and v_a (volts) and b_1 are constants that depend on the temperature and the oxygen partial pressure [1]. The values of v_0, v_a and b_1 can be determined from a nonlinear regression of experimental data. The activation loss is shown in Fig. 4.3(a).

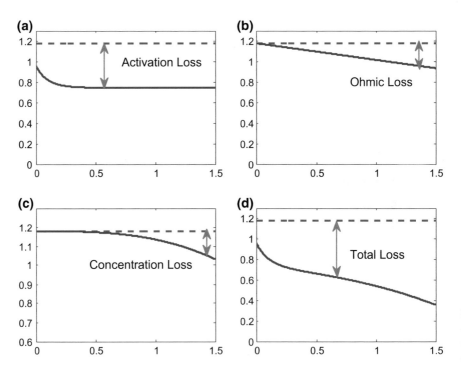

Fig. 4.3 Voltage drops due to different types of losses in FC. **a** Activation losses; **b** Ohmic losses; **c** Concentration losses; **d** Total losses

2. $v_{ohm} = i R_{ohm}$ is due to the electrical resistance of the electrodes, and the resistance to the flow of ions through the electrolyte [6]. R_{ohm} ($\Omega \cdot cm^2$) represents the fuel cell internal electrical resistance. The ohmic loss is shown in Fig. 4.3(b).

3. $v_{conc} = i \left(b_3 \frac{i}{i_{max}} \right)^{b_4}$ results from the drop in concentration of the reactants due to the consumption in the reaction. b_3, b_4 and i_{max} are constants that depend on the temperature and the reactant partial pressures. i_{max} is the current density that generates the abrupt voltage drop. The concentration loss is shown in Fig. 4.3(c).

The stack voltage V_{st} is calculated as the sum of the individual cell voltages,

$$V_{st} = n v_{fc}, \tag{4.4}$$

where n is the number of cells. Thus, the fuel cell power is calculated as

$$P_{st} = I_{st} V_{st}. \tag{4.5}$$

4.3 PEMFC Air-Feed System Model

It is clear from the above discussion that the power generation inside the PEMFC core depends upon several variables and factors, such as oxygen and hydrogen pressures, temperature and various physical parameters of the fuel cell. All these must be controlled rigorously for proper and safe operation of the fuel cell. Hence an operational PEMFC system contains the fuel cell core and a certain number of auxiliary systems for control (Fig. 4.4).

The aim of the air-feed system is to regulate the oxygen quantity in the cathode. It usually consists of an electromechanical air compressor that maintains the required oxygen pressure and mass-flow in the cathode of PEMFC. As mentioned in the

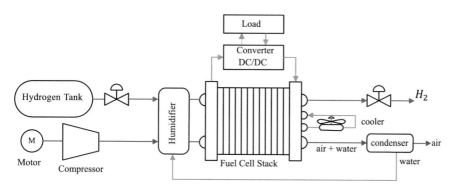

Fig. 4.4 Fuel cell system scheme

introduction, this system can consume up to 30% of the fuel cell power and requires precise control in order to optimize the net power output of the PEMFC system.

This system has been modeled under the following assumptions [11, 12]:

- All gases obey the ideal gas law;
- The temperature of the air inside the cathode is equal to the stack temperature, also is equal to the temperature of the coolant exiting the stack. The stack temperature is well controlled;
- The input reactant flows are humidified in a consistent and rapid way and the high pressure compressed hydrogen is available;
- The water inside the cathode is only in vapor phase and any extra water in liquid phase is removed from the channels;
- The flooding of the gas diffusion layer is neglected;
- The spatial variations are neglected, it is assumed that the flow channel and the gas diffusion layer are lumped into one volume;
- The anode pressure is well controlled to follow the cathode pressure.

4.3.1   Supply Manifold Model

For the supply manifold, the inlet and outlet mass flows are the compressor flow W_{cp} and supply manifold flow $W_{sm,out}$, respectively. The supply manifold model is described by the following equation

$$\frac{dp_{sm}}{dt} = \frac{R_a T_{cp,out}}{V_{sm}} \left(W_{cp} - W_{sm,out} \right), \tag{4.6}$$

where V_{sm} is the supply manifold volume and $T_{cp,out}$ is the temperature of the air leaving the compressor which is calculated as follows:

$$T_{cp,out} = T_{atm} + \frac{T_{atm}}{\eta_{cp}} \left[\left(\frac{p_{sm}}{p_{atm}} \right)^{\frac{\gamma-1}{\gamma}} - 1 \right], \tag{4.7}$$

where η_{cp} is the compressor efficiency (its maximum value is 80%).

The relationship between the flow and the pressure drop can be simplified as a linear nozzle equation since the pressure difference between the supply manifold p_{sm} and the cathode p_{ca} is small:

$$W_{sm,out} = k_{sm,out} \left(p_{sm} - p_{ca} \right), \tag{4.8}$$

where $k_{sm,out}$ is the supply manifold outlet flow constant.

4.3.2 Compressor Model

The air feed system consists of a twin screw compressor and a permanent magnet synchronous (PMSM) motor [8]. In this part, the model of the air compressor is discussed, it is used to provide oxygen to the cathode side of the fuel cell system. The inputs to the model consist of inlet air pressure $p_{cp,in}$, inlet air temperature $T_{cp,in}$ and supply manifold pressure p_{sm}. The inlet air is typically atmospheric and its pressure and temperature are assumed to be $p_{atm} = 1$ atm and $T_{atm} = 25\,°C$, respectively.

The dynamic state in the model, i.e., compressor speed ω_{cp} is given by the following equation

$$\frac{d\omega_{cp}}{dt} = \frac{1}{J_{cp}} \left(\tau_{cm} - \tau_{cp} - f\omega_{cp} \right), \tag{4.9}$$

where J_{cp} is the motor and compressor inertia (kg $\cdot$ m^2), τ_{cm} is the compressor motor torque input (N $\cdot$ m), τ_{cp} is the torque of the compressor (N $\cdot$ m) and f is the friction coefficient.

$$\tau_{cm} = \eta_{cm} k_t I_q,$$

$$\tau_{cp} = \frac{C_p T_{atm}}{\eta_{cp} \omega_{cp}} \left[\left(\frac{p_{sm}}{p_{atm}} \right)^{\frac{\gamma-1}{\gamma}} - 1 \right] W_{cp}, \tag{4.10}$$

where η_{cm} is the motor mechanical efficiency, k_t is the motor constant, I_q is the motor quadratic current, C_p is the specific heat capacity of air (1004 J $\cdot$ kg^{-1} $\cdot$ K^{-1}), γ is the ratio of the specific heats of air ($\gamma = 1.4$) and W_{cp} is the compressor mass flow rate.

The mass flow rate of the twin screw compressor W_{cp} depends on its angular speed ω_{cp}, which is independent of the supply manifold pressure p_{sm} [13]. The following relation is given as

$$W_{cp} = \frac{1}{2\pi} \eta_{v-c} V_{cpr/tr} \rho_a \omega_{cp}, \tag{4.11}$$

Table 4.2 Parameters used in compressor model

Symbols	Parameters	Values
ρ_a	Air density	1.23 kg/m^3
k_t	Motor constant	0.31 N m/A
f	Motor friction	0.00136 V/(rad/s)
J_{cp}	Compressor inertia	671.9 $\times$ 10^{-5} kg m^2
η_{cp}	Compressor efficiency	80%
η_{cm}	Motor mechanical efficiency	98%
$V_{cpr/tr}$	Compressor volume per turn	5 $\times$ 10^{-4} m^3/tr
C_p	Constant pressure specific heat of air	1004 J/(kg K)

where η_{v-c} is the volumetric efficiency, $V_{cpr/tr}$ is the compressed volume per turn and ρ_a is the air density. The parameters used in the compressor model are given in Table 4.2.

The compressor motor power P_{cp} is calculated as follows

$$P_{cp} = \tau_{cm}\omega_{cp}. \tag{4.12}$$

4.3.3 Cathode Flow Model

The thermodynamic properties and mass conservation are used to model the behavior of the air inside the cathode [11]. The dynamics of the oxygen, nitrogen and vapor partial pressures are described by the following equations

$$
\begin{aligned}
\frac{dp_{O_2}}{dt} &= \frac{RT_{fc}}{M_{O_2}V_{ca}} \left(W_{O_2,in} - W_{O_2,out} - W_{O_2,react} \right), \\
\frac{dp_{N_2}}{dt} &= \frac{RT_{fc}}{M_{N_2}V_{ca}} \left(W_{N_2,in} - W_{N_2,out} \right),
\end{aligned}
\tag{4.13}
$$

where $W_{O_2,in}$ is the mass flow rate of oxygen entering the cathode, $W_{O_2,out}$ is the mass flow rate of oxygen leaving the cathode, $W_{O_2,react}$ is the rate of oxygen reacted, $W_{N_2,in}$ is the mass flow rate of nitrogen entering the cathode, $W_{N_2,out}$ is the mass flow rate of nitrogen leaving the cathode, M_{O_2} is the molar mass of oxygen, M_{N_2} is the molar mass of nitrogen and V_{ca} is the cathode volume.

The inlet mass flow rate of oxygen $W_{O_2,in}$ and nitrogen $W_{O_2,out}$ are calculated from the inlet cathode flow $W_{ca,in}$

$$
\begin{aligned}
W_{O_2,in} &= \frac{x_{O_2,atm}}{1 + \omega_{atm}} W_{ca,in}, \\
W_{N_2,in} &= \frac{1 - x_{O_2,atm}}{1 + \omega_{atm}} W_{ca,in},
\end{aligned}
\tag{4.14}
$$

where $x_{O_2,atm}$ is the oxygen mass fraction of the inlet air

$$
x_{O_2,atm} = \frac{y_{O_2,atm} M_{O_2}}{y_{O_2,atm} M_{O_2} + \left(1 - y_{O_2,atm}\right) M_{N_2}}, \tag{4.15}
$$

with the oxygen molar ratio $y_{O_2,atm} = 0.21$ for inlet air and the humidity ratio of inlet air ω_{atm} is defined as

$$
\omega_{atm} = \frac{M_v}{y_{O_2,atm} M_{O_2} + \left(1 - y_{O_2,atm}\right) M_{N_2}} \frac{p_v}{p_{atm} - p_v}, \tag{4.16}
$$

where M_v is the molar mass of vapor and p_v is the vapor pressure.

The vapor partial pressure p_v is determined

$$p_v = \phi_{atm} p_{sat} \left(T_{fc}\right),$$
(4.17)

where ϕ_{atm} is the relative humidify at ambient conditions (its value is set to 0.5), which can be used to describe the relation between the vapor partial pressure and the saturation pressure and $p_{sat} \left(T_{fc}\right)$ is the vapor saturation pressure. The relative humidify ϕ is defined as the ratio of the mole fraction of the water vapor in the mixture (air and water vapor) to the mole fraction of vapor in a saturated mixture at the same temperature and pressure. The value of relative humidify equals 1 means that the mixture is saturated or fully humidified. If there is more water content in the mixture, the extra amount of water condensed into a liquid form will be removed from the channels.

The saturation pressure p_{sat} depends on the temperature and is calculated from the equation given in [9]

$$\log_{10}\left(p_{sat}\right) = -1.69 \times 10^{-10} T_{fc}^4 + 3.85 \times 10^{-17} T_{fc}^3$$
$$- 3.39 \times 10^{-4} T_{fc}^2 + 0.143 T_{fc} - 20.92,$$
(4.18)

where the saturation pressure p_{sat} is in kPa and the temperature T_{fc} is in Kelvin. The cathode inlet flow rate $W_{ca,in}$ is assumed to be the same as the supply manifold outlet flow rate $W_{sm,out}$ (4.8)

$$W_{ca,in} = k_{ca,in} \left(p_{sm} - p_{ca}\right),$$
(4.19)

where the cathode pressure p_{ca} is assumed to be spatially invariant, which is the sum of oxygen, nitrogen and vapor partial pressures,

$$p_{ca} = p_{O_2} + p_{N_2} + p_{sat} \left(T_{fc}\right).$$
(4.20)

The following equations are used to calculate the outlet mass flow rate of oxygen $W_{O_2,out}$ and nitrogen $W_{N_2,out}$ in (4.13)

$$W_{O_2,out} = \frac{x_{O_2,ca}}{1 + \omega_{ca,out}} W_{ca,out},$$
$$W_{N_2,out} = \frac{1 - x_{O_2,ca}}{1 + \omega_{ca,out}} W_{ca,out},$$
(4.21)

where $x_{O_2,ca}, \omega_{ca,out}$ are the oxygen mass fraction, humidity ratio inside the cathode, respectively.

$$x_{O_2,ca} = \frac{y_{O_2,ca} M_{O_2}}{y_{O_2,ca} M_{O_2} + \left(1 - y_{O_2,ca}\right) M_{N_2}},$$

$$\omega_{ca,out} = \frac{M_v}{y_{O_2,ca} M_{O_2} + \left(1 - y_{O_2,ca}\right) M_{N_2}} \frac{p_{sat}}{p_{O_2} + p_{N_2}}. \tag{4.22}$$

Unlike the inlet flow, the oxygen mole fraction of the cathode outlet flow $y_{O_2,ca}$ is not constant since oxygen is reacted, which is calculated as

$$y_{O_2,ca} = \frac{p_{O_2}}{p_{O_2} + p_{N_2}}. \tag{4.23}$$

Using (4.22) and (4.23), Eq. (4.21) can be rewritten as

$$W_{O_2,out} = \frac{M_{O_2} p_{O_2}}{M_{O_2} p_{O_2} + M_{N_2} p_{N_2} + M_v p_{sat}} W_{ca,out},$$

$$W_{N_2,out} = \frac{M_{N_2} p_{N_2}}{M_{O_2} p_{O_2} + M_{N_2} p_{N_2} + M_v p_{sat}} W_{ca,out}. \tag{4.24}$$

The total cathode outlet flow rate $W_{ca,out}$ is calculated by the nozzle flow equation [10],

$$W_{ca,out} = \begin{cases} \frac{C_D A_T}{\sqrt{RT_{fc}}} \left(\frac{p_{atm}}{p_{ca}}\right)^{\frac{1}{\gamma}} \left[\frac{2\gamma}{\gamma-1}\left[1 - \left(\frac{p_{atm}}{p_{ca}}\right)^{\frac{\gamma-1}{\gamma}}\right]\right]^{\frac{1}{2}}, & \text{for } \frac{p_{atm}}{p_{ca}} > \left(\frac{2}{\gamma+1}\right)^{\frac{\gamma}{\gamma-1}}; \\ \frac{C_D A_T p_{ca}}{\sqrt{RT_{fc}}} \gamma^{\frac{1}{2}} \left(\frac{2}{\gamma+1}\right)^{\frac{\gamma+1}{2(\gamma-1)}}, & \text{for } \frac{p_{atm}}{p_{ca}} \leq \left(\frac{2}{\gamma+1}\right)^{\frac{\gamma}{\gamma-1}}. \end{cases} \tag{4.25}$$

The rate of oxygen consumption $W_{O_2,react}$ is a function of the stack current I_{st}, which is calculated using electro-chemistry principles

$$W_{O_2,react} = M_{O_2} \times \frac{n I_{st}}{4F}, \tag{4.26}$$

where F is the Faraday number. Oxygen excess ratio in the cathode λ_{O_2} is defined as the ratio between the amount of oxygen supplied by the compressor and the oxygen reacted in the FC stack. This value is considered as a performance variable of the system since it determines the safety of the fuel cell.

$$\lambda_{O_2} = \frac{W_{O_2,in}}{W_{O_2,react}}. \tag{4.27}$$

4.3.4 Dynamic Model with Four States

In view of Eqs. (4.6), (4.9), and (4.13), the nonlinear model of the fuel cell is completed with the following four states [10]

$$x = \begin{bmatrix} p_{O_2} & p_{N_2} & \omega_{cp} & p_{sm} \end{bmatrix}^{\mathrm{T}}. \tag{4.28}$$

The four state dynamic model is written as follows

$$\dot{x}_1 = b_1(x_4 - x_1 - x_2 - b_2) - \frac{b_3 x_1 W_{ca,out}}{b_4 x_1 + b_5 x_2 + b_6} - b_7 I_{st},$$

$$\dot{x}_2 = b_8(x_4 - x_1 - x_2 - b_2) - \frac{b_3 x_2 W_{ca,out}}{b_4 x_1 + b_5 x_2 + b_6},$$

$$\dot{x}_3 = -b_9 x_3 - \frac{b_{10}}{x_3} \left[\left(\frac{x_4}{b_{11}} \right)^{b_{12}} - 1 \right] W_{cp} + b_{13} u, \tag{4.29}$$

$$\dot{x}_4 = b_{14} \left[1 + b_{15} \left[\left(\frac{x_4}{b_{11}} \right)^{b_{12}} - 1 \right] \right] \times \left[W_{cp} - b_{16}(x_4 - x_1 - x_2 - b_2) \right].$$

The stack current I_{st} is traditionally considered as a disturbance and the control input u is the motor's quadratic current. The outputs of the system are

$$y = \begin{bmatrix} y_1 & y_2 & y_3 \end{bmatrix}^{\mathrm{T}} = \begin{bmatrix} p_{sm} & W_{cp} & V_{st} \end{bmatrix}^{\mathrm{T}}, \tag{4.30}$$

where the stack voltage V_{st}, the supply manifold pressure p_{sm} and the compressor air flow rate W_{cp} are given in Eqs. (4.4), (4.6), and (4.11), respectively. The system performance variables are defined as

$$z = \begin{bmatrix} z_1 & z_2 \end{bmatrix}^{\mathrm{T}} = \begin{bmatrix} P_{net} & \lambda_{O_2} \end{bmatrix}^{\mathrm{T}}, \tag{4.31}$$

where P_{net} is the fuel cell net power and λ_{O_2} is the oxygen excess ratio.

The fuel cell net power P_{net} is the difference between the power produced by the stack P_{st} and the power consumed by the compressor. Thus, the net power can be expressed as

$$P_{net} = P_{st} - P_{cp}, \tag{4.32}$$

where the stack power P_{st} and the compressor power P_{cp} are given in Eqs. (4.5) and (4.12). The oxygen excess ratio λ_{O_2} is defined as the ratio between the oxygen entering the cathode $W_{O_2,in}$ and the oxygen reacting in the fuel cell stack $W_{O_2,react}$

$$\lambda_{O_2} = \frac{W_{O_2,in}}{W_{O_2,react}} = \frac{b_{18}(p_{sm} - p_{ca})}{b_{19} I_{st}}. \tag{4.33}$$

Due to the reasons of safety and high efficiency, it is typical to operate the stacks with this value equals 2 during step changes of current demand [11]. It should be noted that positive deviations of λ_{O_2} above 2 imply lower efficiency, since excess oxygen supplied into the cathode will cause power waste and negative deviations increase the probability of the starvation phenomenon. The parameters b_i, $i \in \{1, \cdots, 19\}$ are defined in Table 4.3.

Table 4.3 Parameter definition

$b_1 = \frac{RT_{fc}k_{ca,in}}{M_{O_2}V_{ca}}\left(\frac{x_{O_2,atm}}{1+\omega_{ca,in}}\right)$	$b_2 = p_{sat}$
$b_3 = \frac{RT_{fc}}{V_{ca}}$	$b_4 = M_{O_2}$
$b_5 = M_{N_2}$	$b_6 = M_v p_{sat}$
$b_7 = \frac{nRT_{fc}}{4FV_{ca}}$	$b_8 = \frac{RT_{fc}k_{ca,in}}{M_{N_2}V_{ca}}\left(\frac{1-x_{O_2,atm}}{1+\omega_{atm}}\right)$
$b_9 = \frac{f}{J_{cp}}$	$b_{10} = \frac{C_p T_{atm}}{J_{cp}\eta_{cp}}$
$b_{11} = p_{atm}$	$b_{12} = \frac{\gamma-1}{\gamma}$
$b_{13} = \frac{\eta_{cm}k_t}{J_{cp}}$	$b_{14} = \frac{RT_{atm}}{M_a V_{sm}}$
$b_{15} = \frac{1}{\eta_{cp}}$	$b_{16} = k_{ca,in}$
$b_{17} = \frac{1}{2\pi}\eta_{v-c}V_{cpr/tr}\rho_a$	$b_{18} = k_{ca,in}\frac{x_{O_2,atm}}{1+\omega_{ca,in}}$
$b_{19} = \frac{nM_{O_2}}{4F}$	

4.4   Experimental Validation

In this section, the experimental validation is performed through the HIL test bench shown in Fig. 4.5. It consists of a physical air-feed system, based on a commercial twin screw compressor and a real-time PEMFC emulator [7, 8]. The complete architecture

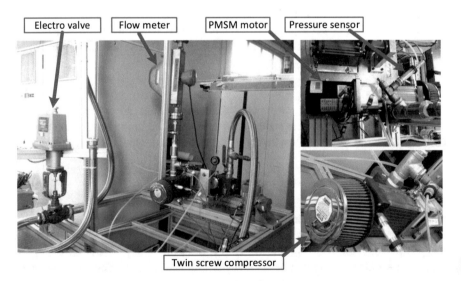

Fig. 4.5   Test bench [8]

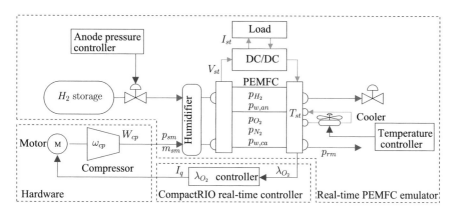

Fig. 4.6 Scheme of HIL system used in the experiments

of the experimental system is shown in Fig. 4.6. The test bench is controlled by National Instruments CompactRIO real-time controller and data acquisition system, using a sampling frequency of 1 kHz.

The objective of the air-feed system is to provide sufficient quantity of oxygen to the PEMFC cathode, keeping the oxygen excess ratio (λ_{O_2}) at its optimum value. The twin screw compressor of the PEMFC air-feed system has a flow rate margin 0–0.1 kg/s at a maximum velocity of 12000 RPM, it consists of two helical rotors which are coupled directly to its motor. Air intake is at the opposite side of the mechanical transmission and the output pressure is regulated by a servo valve. The compressor is driven by a PMSM through an inverter controlled by 3-phase currents I_a, I_b, I_c. The system is controlled by a robust sub-optimal SOSM controller presented in [8].

4.4.1 PEMFC Emulator

The PEMFC emulator (Fig. 4.6) has been designed using experimental data obtained from a 33 kW PEMFC unit containing 90 cells in series. It has been developed on the FPGA platform of CompactRIO running at a clock frequency of 40 MHz and an over all loop frequency of 10 kHz. The nominal value of the PEMFC system parameters obtained from the PEMFC unit (as modeled in the emulator) are given in Table 4.4. The emulator parameters can be varied through software, in order to perform stability and robustness tests. The advantage of this emulator is that it provides an economical and safer alternative for real-time experiments on auxiliary fuel cell system, that does not require excessive hydrogen consumption.

The PEMFC emulator runs a detailed 9th order dynamic PEMFC model, comprising of the following state vector

$$x = \begin{bmatrix} p_{O_2}, & p_{H_2}, & p_{N_2}, & p_{sm}, & m_{sm}, & p_{v,an}, & p_{v,ca}, & p_{rm}, & T_{st} \end{bmatrix}^{\mathrm{T}}, \qquad (4.34)$$

Table 4.4 Emulated PEMFC system parameters [8]

Symbols	Parameters	Values
n	Number of cells in fuel cell stack	90
R	Universal gas constant	8.314 J/(mol K)
R_a	Air gas constant	286.9 J/(kg K)
p_{atm}	Atmospheric pressure	1.01325×10^5 Pa
T_{atm}	Atmospheric temperature	298.15 K
A_{fc}	Active area	800 cm^2
F	Faraday constant	96485 C/mol
M_a	Air molar mass	28.9644×10^{-3} kg/mol
M_{O_2}	Oxygen molar mass	32×10^{-3} kg/mol
M_{N_2}	Nitrogen molar mass	28×10^{-3} kg/mol
M_{H_2}	Hydrogen molar mass	2×10^{-3} kg/mol
M_v	Vapor molar mass	18.02×10^{-3} kg/mol
C_D	Discharge of the nozzle	0.0038
A_T	Operating area of the nozzle	0.00138 m^2
γ	Ratio of specific heats of air	1.4
J_{cp}	Compressor inertia	6719 mg $\cdot$ m^2
f	Motor friction	1.36 mV/(rad/s)
k_t	Motor constant	0.31 N m/A
C_p	Specific heat of air	1004 J/(Kg $\cdot$ K)
η_{cm}	Motor mechanical efficiency	98%
V_{ca}	Cathode volume	0.0015 m^3
V_{sm}	Supply manifold volume	0.003 m^3
V_{an}	Anode volume	0.0005 m^3
$V_{cpr/tr}$	Compressor volume per turn	$5 \cdot 10^{-4}$ m^3/tr
$k_{ca,in}$	Cathode inlet orifice constant	0.3629×10^{-5} kg/(Pa s)
$k_{an,in}$	Anode inlet orifice constant	0.21×10^{-5} kg/(Pa s)
ρ_a	Air density	1.23 kg/m^3
$x_{O_2,ca,in}$	Oxygen mass fraction	0.23

where p_{H_2} is hydrogen pressure in the anode, m_{sm} is mass of air in supply manifold, $p_{v,an}$ is vapor partial pressure in the anode, $p_{v,ca}$ is vapor partial pressure in the cathode, p_{rm} is pressure of return manifold and T_{st} is the stack temperature. This dynamic model is based on Pukrushpan et al. [11], with the added temperature model described by a lumped thermal model [11]

$$
\begin{aligned}
m_{st} C_{p_{st}} \frac{dT_{st}}{dt} &= \dot{Q}_{sou} - W_c C_{p_c} \left(T_{st} - T_{c,in} \right), \\
\dot{Q}_{sou} &= I_{st} \left(-\frac{T \Delta s}{4F} + v_{act} + I_{st} R_{ohm} \right),
\end{aligned}
\tag{4.35}
$$

where m_{st} is the heat mass of the stack, $C_{p_{st}}$ and C_{p_c} are the specific heat, W_c is the coolant flow rate considered as a control variable, $T_{c,in}$ is the coolant temperature at the stack inlet and $\dot{Q}_{sou}$ is the internal energy source. The latter is calculated as a function of the stack current, temperature, electrical resistance of stack layers R_{ohm}, Faraday's number F and the entropy change Δs. The physical parameters were obtained through extensive experimentation.

The PEMFC works under several safety constraints, the most important of which are the anode and cathode pressure difference (to be kept minimum) and stack temperature (to be regulated). These constraints require additional control in real fuel cells, which have been replicated in the emulator as well. An internal controller is implemented in the emulator to ensure that the anode pressure follows the cathode pressure at all times. The temperature is also controlled and its value can be set through software. This added characteristic is a key feature of the emulator, as compared with other works [2, 4, 11, 14], since the heat produced in real fuel cells (due to heat produced by irreversible energy occurring in the chemical reactions and Joules losses) is a major factor that causes large parametric variations in the system.

4.4.2 Validation of 4th Order Model with PEMFC Emulator

The experimental stack voltage, the predicted stack voltage by the 4-th order model and the predicted stack voltage by the PEMFC emulator, as well as the experimental data obtained from the 33 kW PEMFC stack are shown and compared in Fig. 4.7. From this figure, it can be concluded that the 4-th order model replicates the

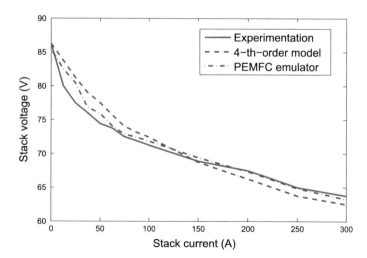

Fig. 4.7 Stack voltage response

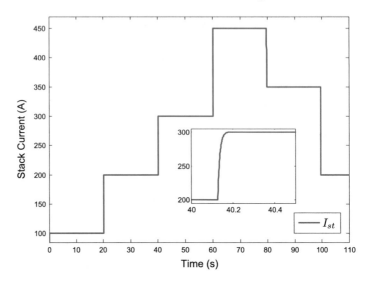

Fig. 4.8 Stack current under load variation

characteristics of the Fuel Cell with sufficient precision with a relative error less than
2.5% ($\pm$1.6 volts).

To demonstrate the FC model characteristics, the stack current shown in Fig. 4.8
was varied between 100 and 450 A during the tests and a static feed-forward controller
is used to control the compressor voltage so that the oxygen excess ratio maintains
near 2.

It can be seen from the Fig. 4.9(b) that a step increase in the load current (i.e. t
= 20 s) causes a drop in the oxygen excess ratio. The performance variables (P_{net},
λ_{O_2}) in Fig. 4.9 and the states variables (p_{O_2}, p_{N_2}, ω_{cp}, p_{sm}) in Figs. 4.10 and 4.11
show that the four states model matches well with the emulator outputs.

4.5   Conclusion

In this chapter, we introduced different types of fuel cells, i.e. PEMFC, AFC, PAFC,
SOFC and MCFC. The advantages and disadvantages of these fuel cells were summarized. Then, we focus on the type of PEMFC, which is suitable for automobile
applications due to its high energy density, low working temperature, simple structure and long life. PEMFC is supplied with hydrogen and air at the anode and the
cathode, respectively.

A typical PEMFC system consists of four subsystems: the air feed subsystem, the
hydrogen supply subsystem, the humidifier subsystem and the cooling subsystem.
The detail models of the air feed subsystem including air compressor and fuel cell
stack were given. The studied air feed system consists of a 33 kW PEMFC containing

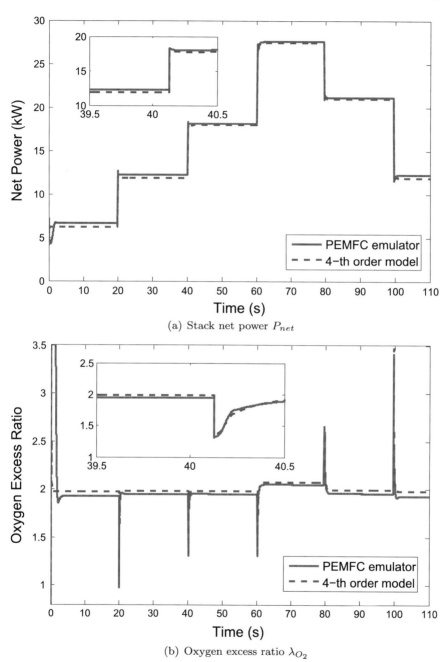

(a) Stack net power P_{net}

(b) Oxygen excess ratio λ_{O_2}

Fig. 4.9 Experimental validation of P_{net} and λ_{O_2}

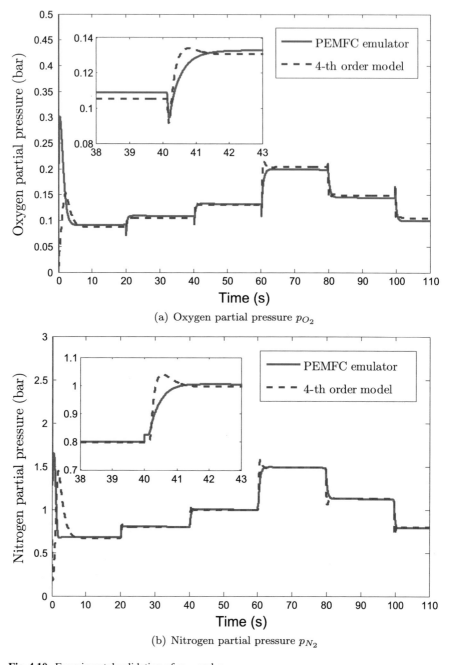

(a) Oxygen partial pressure p_{O_2}

(b) Nitrogen partial pressure p_{N_2}

Fig. 4.10 Experimental validation of p_{O_2} and p_{N_2}

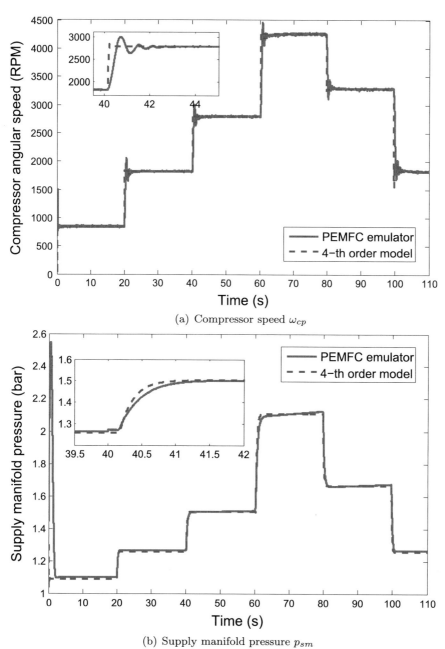

(a) Compressor speed ω_{cp}

(b) Supply manifold pressure p_{sm}

Fig. 4.11 Experimental validation of ω_{cp} and p_{sm}

90 cells in series and a twin screw compressor. Under several practical assumptions, this model is reduced into a four states model while preserving the dynamic behavior. Finally, experimental validation was performed through the HIL test bench which consists of a physical air-feed system, based on a commercial twin screw compressor and a real-time PEMFC emulator. In the following chapter, disturbance-observer based SOSM control will be designed for the PEMFC system based on the proposed model.

References

1. Amphlett, J.C., Baumert, R., Mann, R.F., Peppley, B.A., Roberge, P.R., Harris, T.J.: Performance modeling of the Ballard Mark IV solid polymer electrolyte fuel cell II. Empirical model development. J. Electrochem. Soc. **142**(1), 9–15 (1995)
2. Arce, A., del Real, A., Bordons, C., Ramirez, D.: Real-time implementation of a constrained MPC for efficient airflow control in a pem fuel cell. IEEE Trans. Ind. Electron. **57**(6), 1892–1905 (2010)
3. Grove, W., Egeland, O.: A small voltaic battery of great energy. Philos. Mag. **15**, 287–293 (1839)
4. Gruber, J., Bordons, C., Oliva, A.: Nonlinear MPC for the airflow in a PEM fuel cell using a volterra series model. Control. Eng. Pract. **20**(2), 205–217 (2012)
5. Kroposki, B.: DUIT: Distributed utility integration test. National Renewable Energy Laboratory, Distributed Utility Associates Livermore, California (2003). https://www.nrel.gov/docs/fy03osti/34389.pdf
6. Larminie, J., Dicks, A., McDonald, M.S.: Fuel Cell Systems Explained, vol. 2. Wiley, Chichester (2003)
7. Liu, J., Laghrouche, S., Ahmed, F.S., Wack, M.: PEM fuel cell air-feed system observer design for automotive applications: an adaptive numerical differentiation approach. Int. J. Hydrog. Energy **39**(30), 17210–17221 (2014)
8. Matraji, I., Laghrouche, S., Jemei, S., Wack, M.: Robust control of the PEM fuel cell air-feed system via sub-optimal second order sliding mode. Appl. Energy **104**, 945–957 (2013)
9. Nguyen, T.V., White, R.E.: A water and heat management model for proton-exchange-membrane fuel cells. J. Electrochem. Soc. **140**(8), 2178–2186 (1993)
10. Pukrushpan, J., Peng, H., Stefanopoulou, A.: Control-oriented modeling and analysis of fuel cell reactant flow for automotive fuel cell systems. ASME J. Dyn. Syst., Meas. Control. **126**(1), 14–25 (2004)
11. Pukrushpan, J.T., Stefanopoulou, A.G., Peng, H.: Control of Fuel Cell Power Systems: Principles, Modeling, Analysis and Feedback Design. Springer Science & Business Media (2004)
12. Suh, K.W.: Modeling, analysis and control of fuel cell hybrid power systems. Ph.D. thesis, Department of Mechanical Engineering, The university of Michigan (2006)
13. Talj, R.J., Hissel, D., Ortega, R., Becherif, M., Hilairet, M.: Experimental validation of a PEM fuel-cell reduced-order model and a moto-compressor higher order sliding-mode control. IEEE Trans. Ind. Electron. **57**(6), 1906–1913 (2010)
14. Tekin, M., Hissel, D., Pera, M.C., Kauffmann, J.M.: Energy-management strategy for embedded fuel-cell systems using fuzzy logic. IEEE Trans. Ind. Electron. **54**(1), 595–603 (2007)

Chapter 5
Sliding Mode Control of PEMFC Systems

In this chapter, a model-based robust control is proposed for PEMFC system, based on SOSM algorithm. The control objective is to maximize the fuel cell net power and avoid the oxygen starvation by regulating the oxygen excess ratio to its desired value during fast load variations. The oxygen excess ratio is estimated via an ESO from the measurements of the compressor flow rate, the load current and supply manifold pressure. A HIL test bench which consists of a commercial twin screw air compressor and a real-time fuel cell emulation system, is used to validate the performance of the proposed ESO-based SOSM controller. The experimental results show that the proposed controller is robust and has a good transient performance in the presence of load variations and parametric uncertainties.

5.1 Introduction

Nowadays, fuel cells are widely regarded as a potential alternative energy conversion technology for stationary and mobile applications since they have high efficiency and low environmental pollution [5, 13, 23]. Among various types of fuel cells, polymer electrolyte membrane fuel cells (PEMFCs) are particularly suitable for automotive applications due to low temperature, quick start-up and favorable power-to-weight ratio [11, 34]. However, PEMFCs are multi-input multi-output systems with strong nonlinearities, and thus precise control efforts are required to guarantee the optimal performance. One of the most challenging tasks is related with the fuel cell air flow control, which means to regulate the oxygen quantity in the cathode. On one hand, during sudden load increase, it must ensure that the oxygen partial pressure supplied by the air compressor does not fall below a critical level (the value of oxygen excess ratio is less than one). Otherwise, the so-called *oxygen starvation* phenomenon occurs which will result in hot spots on the surface of the membrane, causing irreversible

© Springer Nature Switzerland AG 2020   83
J. Liu et al., *Sliding Mode Control Methodology in the Applications of Industrial Power Systems*, Studies in Systems, Decision and Control 249,
https://doi.org/10.1007/978-3-030-30655-7_5

damages to the polymeric membranes [8]. On the other hand, it must also reduce the parasitic losses of the air compressor since it can consume up to 30% of the fuel cell power under high load conditions [35]. Therefore, the management of air supplying to the fuel cell is one of the key factors for achieving desired fuel cell operation conditions.

Many model based control strategies have been proposed for the control design of fuel cell air feed systems, ranging from linear state feedback control [26], optimal control [22], model predictive control [9, 32], intelligent control [6, 14] and SMC [8, 23]. One of the main disadvantages of employing the intelligent control method [6] is the requirement of off-line training, which is difficult for real applications because of high computational burden. Furthermore, the neural network/fuzzy rule-based model does not provide explicit structure property of the physical system. As discussed in [14], the critical problem in fuel cell control design is the reconstruction of the oxygen excess ratio value. The accurate regulation of oxygen excess ratio can increase the system efficiency significantly [28]. Unfortunately, this value depends on the oxygen flow entering the cathode, which is unavailable for measurement. Therefore, state observers employed as a replacement for physical sensors to estimate the unavailable variables, are of great interest.

Many observation/estimation strategies have been proposed for the fuel cell systems in the last few years. Several kinds of Kalman filters (KFs) have been applied to the state estimation of the fuel cell systems [26, 33]. It should be noted that these methods are sensitive to external disturbances and modeling errors due to model linearization around pre-defined operating points of the system [2, 15]. Furthermore, the calculation of Jacobian matrix of the fuel cell system is time consuming. In [1], an adaptive observer was designed to estimate the states of PEMFCs. However, it lacks robustness against the fuel cell voltage's measurement noise and is dependent on the system parameters. A finite-time convergent state observer based on HOSM was proposed for the PEMFCs, which allows for the finite-time estimation of the fuel cell states [23, 27]. Consequently, the separation principle is automatically satisfied, which means that the controller and the observer can be separately designed. This method relies on the calculation of the complex observability matrices which are highly nonlinear and state dependent, therefore difficult in real-time implementation. More recently, ESO has attracted great attention because it requires the least amount of plant information. Unlike traditional observers, such as Luenberger observer, high-gain observer and unknown input observer, ESO regards the disturbances of the system as a new system state which is conceived to estimate not only the external disturbance but also plant dynamics.

In this chapter, an ESO-based cascade controller is designed for regulating the oxygen excess ratio of the PEMFC air compression system to its desired value, using sliding mode technique. SMC is known for its finite time convergence, high accuracy and insensitiveness to uncertainties and disturbance [3, 12, 19]. These make it an effective method to deal with the nonlinear behavior of the considered system. The proposed controller is designed in a cascade structure because the relative degree between the oxygen excess ratio and the compressor quadrature current is two. It consists of two control loops, i.e., ESO-based oxygen excess ratio regulation loop (outer loop) and compressor flow rate control loop (inner loop). The oxygen excess ratio is reconstructed by an ESO-based on the measurements of supply manifold pressure and compressor flow rate. The outer control loop, which uses the estimated oxygen excess ratio, provides the compressor flow rate reference for the inner loop based on the STA. The STA is an unique absolutely continuous sliding mode algorithm among the SOSM algorithms. Therefore, without any introduction of low pass filters [19], it is less affected by the chattering phenomenon. For the inner control loop, a simple SMC law consists of a linear term and a switching term is adopted, ensuring a fast response of the control scheme. The problems related to practical implementation of the proposed ESO-based cascade control on an HIL test bench has been resolved, and are the main focus of this paper. Experimental results are presented and the robustness of the proposed algorithm against parametric variation has been validated.

5.2   Problem Formulation

A schematic diagram of a PEMFC system is shown in Fig. 5.1. The considered PEMFC system consists of four major subsystems: the air feed subsystem that feeds the cathode by oxygen, the hydrogen supply subsystem that feeds the anode side with hydrogen, the humidify subsystem and the cooling subsystem that maintain the humidity and the temperature of the fuel cells respectively. The air feed subsystem usually consists of an electromechanical air compressor that maintains the required oxygen pressure and mass-flow in the cathode of PEMFC.

There are a number of models of fuel cells and air feed systems in the literature, such as the nine-state model proposed by Pukrushpan et al. [26], one dimensional power density model proposed by Meidanshahi and Karimi [21], and multi-dimensional models developed in [4]. Although these works are useful for analytical purposes, they require large quantity of calculations. Thus, it is necessary to carefully consider the model since some weakly coupled states can lead to a problem of observability of the states [33]. Suh [30] has reduced the nine-state model of Pukrushpan et al. [26] into a four-state model, while preserving the dynamic behavior. The advantage of this model is that it has been validated in a wide operating range of a fuel cell by experimental results.

By taking the dynamics of the vapor partial pressure into account, a 5-th order model of the PEMFC air feed system is considered in this work. This model facilitates both the control and observer design of the fuel cell air feed system in real-time. Some assumptions are used to derive a more accurate model of the PEMFC system which includes the cathode partial pressure dynamics, the air supply manifold dynamics and the compressor dynamics [27, 30]. Mainly, the temperature and the humidity of the air at the inlet of the fuel cell stack are assumed to be well controlled. The high pressure compressed hydrogen is available and is well controlled to follow the cathode pressure. The current dynamics of the motor which directly drive the

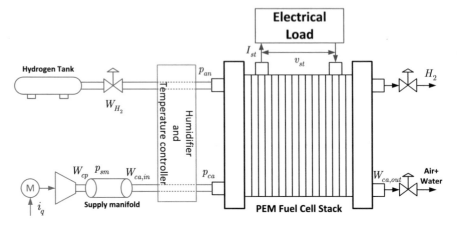

Fig. 5.1 Fuel cell system scheme

compressor are negligible due to their small time constant as compared to the mechanical dynamics.

Some important assumptions have been considered for the system modelling [27, 30]:

- The temperature and the humidity of the air at the inlet of the fuel cell stack are assumed to be well controlled. This assumption is reasonable due to the fact the dynamics of the temperature and the humidity are very slow.
- The high pressure compressed hydrogen is available and is well controlled to follow the cathode pressure.
- The water inside the cathode is only in vapor phase and any extra water in liquid phase is removed from the channels.
- The spatial variations are neglected, it is assumed that the gases and reactions are uniformly distributed in the cell.
- The current dynamics of the motor which directly drives the compressor is negligible due to its small time constant as compared to the mechanical dynamics.

5.2.1   The Vapor Partial Pressures Dynamics

The dynamics of the vapor partial pressures are described by the following equation

$$\frac{dp_{v,ca}}{dt} = \frac{R_v T_{fc}}{V_{ca}} \left(W_{v,ca,in} - W_{v,ca,out} + W_{v,ca,gen} + W_{v,mem} \right), \qquad (5.1)$$

where $W_{v,ca,in}$ and $W_{v,ca,out}$ are the inlet and outlet flow rates of vapor, respectively, $W_{v,ca,gen}$ is the generated vapor flow and $W_{v,mem}$ is the mass flow rate across the membrane. The detail calculation of these values can be found in [26].

5.2.2   State Space Representation of PEMFC System

In view of Eqs. (4.6, 4.9, 4.13, 5.1) and define the state vector,

$$x = [x_1, x_2, x_3, x_4, x_5]^T = [w_{cp}, p_{sm}, p_{O_2}, p_{N_2}, p_{v,ca}]^T. \qquad (5.2)$$

Denote $h_2 := W_{cp}$ and $\chi := p_{ca} = x_3 + x_4 + x_5$. Thus, the resulting nonlinear model, suitable for control design, can be written in the following form

$$\dot{x}_1 = -c_1 x_1 - \frac{c_2}{x_1} \left[\left(\frac{x_2}{c_3} \right)^{c_4} - 1 \right] h_2 + c_5 u, \qquad (5.3)$$

$$\dot{x}_2 = c_6 \left\{ 1 + c_7 \left[\left(\frac{x_2}{c_3} \right)^{c_4} - 1 \right] \right\} [h_2 - c_8 (x_2 - \chi)], \qquad (5.4)$$

$$\dot{x}_3 = c_9 (x_2 - \chi) - \frac{c_{10} x_3 \varphi (\chi)}{c_{11} x_3 + c_{12} x_4 + c_{13} x_5} - c_{14} \zeta, \qquad (5.5)$$

$$\dot{x}_4 = c_{15} (x_2 - \chi) - \frac{c_{10} x_4 \varphi (\chi)}{c_{11} x_3 + c_{12} x_4 + c_{13} x_5}, \qquad (5.6)$$

$$\dot{x}_5 = c_{16} (x_2 - \chi) - \frac{c_{10} x_5 \varphi (\chi)}{c_{11} x_3 + c_{12} x_4 + c_{13} x_5} + 2 c_{14} \zeta + c_{17}, \qquad (5.7)$$

where $u := i_q$ is the control input, $\zeta := I_{st}$ is considered as a measurable disturbance, and the function $\varphi (\chi)$ represents the flow rate at the cathode exit which depends on χ. The parameters c_i, $i \in \{1, \ldots, 21\}$, are given in Table 5.1.

As mentioned in [25], the vector of measurable outputs is given as follows:

$$y = \begin{bmatrix} y_1 \\ y_2 \\ y_3 \end{bmatrix} = \begin{bmatrix} h_1 (x_3, x_4) \\ h_2 (x_1) \\ x_2 \end{bmatrix}, \qquad (5.8)$$

where $y_1 := h_1 = V_{st}$, $y_2 := h_2 = W_{cp}$ and $y_3 := x_2 = p_{sm}$. It should be noted that all physical quantities are measured in the SI system of units.

5.2.3   Control Objectives

As the PEMFC system works as an autonomous power plant in automobiles, the compressor motor is also powered by the PEMFC. Therefore, the net power of the system can be expressed as (consumption by the other auxiliary systems is negligible):

$$P_{net} = P_{st} - P_{cp},$$

Table 5.1 Parameters
defined in system (5.3)–(5.7)

$c_1 = \frac{\eta_{cm} k_t k_v}{J_{cp} R_{cm}}$	$c_2 = \frac{C_p T_{atm}}{J_{cp} \eta_{cp}}$
$c_3 = p_{atm}$	$c_4 = \frac{\gamma - 1}{\gamma}$
$c_5 = \frac{\eta_{cm} k_t}{J_{cp}}$	$c_6 = \frac{R T_{atm}}{M_a V_{sm}}$
$c_7 = \frac{1}{\eta_{cp}}$	$c_8 = k_{ca,in}$
$c_9 = \frac{R T_{st} k_{ca,in}}{M_{O_2} V_{ca}} \frac{x_{O_2,atm}}{1 + \omega_{ca,in}}$	$c_{10} = \frac{R T_{st}}{V_{ca}}$
$c_{19} = M_{O_2}$	$c_{12} = M_{N_2}$
$c_{13} = M_{v,ca}$	$c_{14} = \frac{n R T_{fc}}{4 F V_{ca}}$
$c_{15} = \frac{R T_{st} k_{ca,in}}{M_{N_2} V_{ca}} \frac{1 - x_{O_2,atm}}{1 + \omega_{atm}}$	$c_{16} = \frac{R_v T_{st} \omega_{atm}}{1 + \omega_{atm}} k_{ca,in}$
$c_{17} = \frac{R T_{st}}{V_{ca}} N_{v,membr} A_{fc} n$	$c_{18} = k_{ca,in} \frac{x_{O_2,ca,in}}{1 + \omega_{atm}}$
$c_{19} = \frac{n M_{O_2}}{4F}$	$c_{20} = \frac{c_{18}}{c_{19} \zeta}$
$c_{21} = \frac{1}{2\pi} \eta_{v-c} V_{cpr/tr} \rho_a$	

where P_{st} and P_{cp} are the powers produced by the fuel cell and the air-feed system, respectively.

Experimental studies have shown that the air-feed system can consume up to 30% of the fuel-cell power under high load conditions [32]. Therefore, it needs to be operated at its optimal point, at which it supplies just sufficient oxygen necessary for the hydrogen and oxygen reaction. As already discussed in [26], accomplishing the power optimization is equivalent to maintain the oxygen excess ratio in an optimal value $\lambda_{O_2}^*$.

The oxygen excess ratio λ_{O_2} is defined as [26, 28]:

$$\lambda_{O_2} = \frac{W_{O_2,in}}{W_{O_2,react}} = \frac{c_{18}}{c_{19} \zeta} (p_{sm} - \chi), \tag{5.9}$$

where $W_{O_2,in}$ is the oxygen partial flow in the cathode and $W_{O_2,react}$ is the reacted oxygen flow. Therefore, the main control objective for the fuel cell air-fed system is to force the oxygen excess ratio λ_{O_2} track $\lambda_{O_2}^*$. Given that the relative degree between the oxygen excess ratio and control input is 2, an external control loop which regulates the compressor flow rate $h_2(x_1)$ to its desired value $h_2^{ref}(x_1)$ needs to be considered. The desired compressor flow rate is calculated in such a way that $\lambda_{O_2} \rightarrow \lambda_{O_2}^*$.

Remark 5.1 High oxygen excess ratio, and thus high oxygen partial pressure, improves the net power P_{net} and the stack power P_{st}. However, above an optimum $\lambda_{O_2}^*$

level, further increase will cause excessive increase of compressor power and thus deteriorate the system net power. For different load current I_{st} drawn from the fuel cell system, there is an optimum $\lambda_{O_2}^*$ lies between 2 and 2.5 where the maximized net power is achieved.

5.3 Control Design

5.3.1 Oxygen Excess Ratio Control

In view of (5.9), the main obstacle of calculating the oxygen excess ratio λ_{O_2} is the unavailability of the cathode pressure χ which is difficult to be measured. Furthermore, it is not always possible to use sensors for measurements, either due to prohibitive costs of the sensing technology or because the quantity is not directly measurable, especially in the conditions of humidified gas streams inside the fuel cell stack. In these cases, state observers serve as a replacement for physical sensors, for obtaining the unavailable quantities, are of great interest [7].

Hence, we propose an ESO which provides satisfactory level of accuracy and fast convergence for reconstructing the stoichiometry λ_{O_2}. Then, the estimated value is used in the output feedback controller design which takes a cascade control structure (see Fig. 5.2). The proposed control structure consists of two control loops: outer loop and inner loop. The outer loop generates a reference compressor flow rate based on the oxygen excess ratio error. The inner loop computes the control input u based on the error information between the reference compressor flow rate and measured value. The super-twisting (ST) sliding mode algorithm is employed in both outer and inner control loops.

Extended State Observer Design

The Eq. (5.4) can be rewritten as follows:

$$\dot{x}_2 = f_1(x_2)(h_2 - c_8 x_2),\qquad(5.10)$$

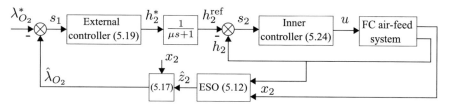

Fig. 5.2 Block diagram of the oxygen excess ratio control system

where $f_1(x_2) = c_6 \left\{ 1 + c_7 \left[\left(\frac{x_2}{c_3} \right)^{c_4} - 1 \right] \right\}$ and $d(t) = c_8 f_1(x_2) \chi$.

Remark 5.2 Given that x_2 is a physical variable and the coefficients c_3, c_4, c_6 and c_7 are all positive, the term $\left(\frac{x_2}{c_3} \right)^{c_4} - 1 > 0$ is assured in the operation range, which means $f_1(x_2)$ is also positive. Thus, it is reasonable to consider that $f_1(x_2)$ is a bounded invertible function in the operation range $x_2 \in X_2$.

Denote $z_1 := x_2$ and $z_2 := d(t)$. Assume that $f_1(x_2)$ and χ are both differentiable, it follows $\dot{z}_2 = h(t)$ with $h(t)$ the variation rate of system uncertainty and disturbance $d(t)$. Then, (5.10) can be rearranged as,

$$\begin{aligned} \dot{z}_1 &= f_2(z_1, h_2) + z_2, \\ \dot{z}_2 &= h(t), \end{aligned} \tag{5.11}$$

in which $f_2(z_1, h_2) = f_1(z_1)(h_2 - c_8 z_1)$. A linear ESO is designed with the following structure [10, 31]:

$$\begin{aligned} \dot{\hat{z}}_1 &= f_2(z_1, h_2) + \hat{z}_2 + \beta_1 \left(z_1 - \hat{z}_1 \right), \\ \dot{\hat{z}}_2 &= \beta_2 \left(z_1 - \hat{z}_1 \right), \end{aligned} \tag{5.12}$$

where $\hat{z} = \left[\hat{z}_1, \hat{z}_2 \right]^{\mathrm{T}} \in \mathcal{R}^2$, the designing gains β_1 and β_2 are chosen such that the polynomial $\lambda^2 + \beta_1 \lambda + \beta_2$ is Hurwitz stable. Therefore, its natural frequency ω_n and damping ratio ξ are:

$$\omega_n = \sqrt{\beta_2}, \ \xi = \frac{\beta_1}{2\sqrt{\beta_2}}.$$

Denote the observation errors $\tilde{z}_1 = z_1 - \hat{z}_1$ and $\tilde{z}_2 = z_2 - \hat{z}_2$, the error dynamics are given by,

$$\begin{aligned} \dot{\tilde{z}}_1 &= -\beta_1 \tilde{z}_1 + \tilde{z}_2, \tag{5.13} \\ \dot{\tilde{z}}_2 &= -\beta_2 \tilde{z}_1 + h(t). \tag{5.14} \end{aligned}$$

The system (5.13) can be written in state space form as follows:

$$\dot{\tilde{z}}_{12} = A\tilde{z}_{12} + Bh(t), \tag{5.15}$$

where $\tilde{z}_{12} = \begin{bmatrix} \tilde{z}_1 \\ \tilde{z}_2 \end{bmatrix}$, $A = \begin{bmatrix} -\beta_1 & 1 \\ -\beta_2 & 0 \end{bmatrix}$ and $B = \begin{bmatrix} 0 \\ 1 \end{bmatrix}$.

Lemma 5.1 *Assuming that $h(t)$ is bounded, there exist a constant $\delta > 0$ and a finite time $T_1 > 0$ such that the trajectories of the system (5.13) are bounded, $\|\tilde{z}_{12}\| \leq \delta$, $\forall t \geq T_1 > 0$ and δ is dependent on the initial condition of $\tilde{z}_{12}$ and the upper boundary of $h(t)$.*

Proof Under the assumption of the boundedness of $h(t)$, the bounded-input-bounded-output stability for the system (5.13) is ensured. The detailed proof can be obtained directly from the works [31, 36]. ∎

From Remark 5.2, $f_1(x_2)$ is a bounded invertible function in the operation range. Then, it follows from Lemma 5.1 that the estimate of χ can be constructed as,

$$\hat{\chi} = c_8^{-1} f_1^{-1}(x_2) \hat{z}_2, \quad \text{for } f_1(x_2) \neq 0. \tag{5.16}$$

Remark 5.3 Since $h(t)$ is unknown in the error dynamics (5.14), it should be noted that the estimation of χ in (5.16) is not asymptotic stable in the case when $h(t)$ is not negligible. For the system having fast varying disturbances/uncertainties, i.e., $h(t)$ is large, the performance of ESO presented above does not offer satisfactory results. To deal with this issue, the concept of higher order ESO can be employed. The idea is simple, if $h(t)$ is not negligible but its derivative $\dot{h}(t)$ is negligible, then a second order ESO can be used. Therefore, in application, one can design a r-th order ESO if $h^{(r-1)}(t)$ is negligible [31].

Remark 5.4 The observer gains β_1 and β_2 need to be determined, which decides the bandwidth of the ESO. For instance, we can choose $\beta_1 = 2\beta > 0$ and $\beta_2 = \beta^2$, in view of (5.15), the larger value β is, the more accurate estimation is achieved. However, this increases the noise sensitivity due to the augment of the bandwidth. More generally speaking, the function $h(t)$ represents the rate of change in the physical world. If this value is quite large, the physical variables will change very rapidly. In this case, the observer bandwidth needs to be sufficiently large for an accurate estimate of z_2. Therefore, the selection of β should balance the estimation performance and the noise tolerance.

Outer Loop

From the result of Lemma 5.1, the oxygen excess ratio can be reconstructed as,

$$\hat{\lambda}_{O_2} = \frac{c_{18}}{c_{19}\zeta}(x_2 - \hat{\chi}). \tag{5.17}$$

The sliding variable is defined as follows:

$$s_1(t) = \hat{\lambda}_{O_2} - \lambda_{O_2}^*,$$

where $\lambda_{O_2}^*$ is the reference of oxygen excess ratio, whose value is between 2.0 and 2.5. A variable setting for the reference of λ_{O_2} is given in order to not only obtain the maximum efficiency from the fuel cell but also avoid the dangers of oxygen starvation in the presence of fast load variations [8, 16].

Take the first time derivative of the sliding variable $s_1(t)$:

$$\dot{s}_1(t) = c_{20}\left[f_1(x_2)(h_2 - c_8x_2 + c_8\chi) - \dot{\hat{\chi}}\right]$$
$$= g_1(x_2)h_2 + \varphi_1\left(x_2, \chi, \dot{\hat{\chi}}\right), \tag{5.18}$$

where $g_1(x_2) = g_{10} + \Delta g_{10}$, and $g_{10} > 0$ is a known function based on nominal parameters while Δg_{10} represents parametric uncertainties, $\varphi_1\left(x_2, \chi, \dot{\hat{\chi}}\right) = \hat{\varphi}_1\left(x_2, \hat{\chi}, \dot{\hat{\chi}}\right) + \Delta\varphi_1$. It is assumed that $0 < \frac{|\Delta g_{10}|}{g_{10}} \le b_1 < 1$ and $\Delta\varphi_1$ is assumed to be bounded and with first derivative bounded [29]. Both g_1 and φ_1 are smooth enough in the operation range.

Considering the convergence and stability of the system, the STA can be employed for compressor mass flow control in order to force $s_1(t)$ and $\dot{s}_1(t)$ to zero in finite time. The desired compressor flow rate h_2^* is calculated as follows:

$$h_2^* = \frac{1}{g_{10}}\left[-\sigma(s_1) - \hat{\varphi}_1\left(x_2, \hat{\chi}, \dot{\hat{\chi}}\right)\right], \tag{5.19}$$

where $\sigma(s_1) = \lambda_1|s_1|^{\frac{1}{2}}\text{sign}(s_1) + \alpha_1\int_0^t \text{sign}(s_1)d\tau$ is the STA, λ_1 and α_1 are design parameters. Substitute (5.19) into (5.18), the sliding manifold can be rewritten as follows:

$$\dot{s}_1(t) = -\left(1 + \frac{\Delta g_{10}}{g_{10}}\right)\sigma(s_1) + \Delta\varphi_1 - \frac{\Delta g_{10}}{g_{10}}\hat{\varphi}_1$$
$$= -\gamma_1\sigma(s_1) + \phi_1.$$

Through the application of the extended state observer, it can yield that the term $\phi_1 = \Delta\varphi_1 - \frac{\Delta g_{10}}{g_{10}}\hat{\varphi}_1$ is bounded and with first derivative bounded, i.e., $\frac{d}{dt}|\phi_1| \le \Phi_1$, for some positive value Φ_1. The sufficient conditions for the finite time convergence to the sliding manifold $s_1(t) = \dot{s}_1(t) = 0$ are [18]:

$$\alpha_1 > \frac{\Phi_1}{1 - b_1}, \quad \lambda_1^2 \ge \frac{4\Phi_1}{(1 - b_1)^2}\frac{\alpha_1 + \Phi_1}{\alpha_1 - \Phi_1}. \tag{5.20}$$

The boundary values of these functions need to be determined a-priori through experimentation. Then, the controller gains α_1 and λ_1 can be selected according to (5.20). In this study, we have found the following boundary values as $\Phi_1 = 4.2 \times 10^3$, $b_1 = 0.4$. Accordingly, the gains are tuned as $\alpha_1 = 7.5 \times 10^3$ and $\lambda_1 = 425$.

The controller (5.19) is followed by a linear first order filter which is given as follows [24]:

$$\mu\frac{d}{dt}h_2^{\text{ref}} = h_2^* - h_2^{\text{ref}}. \tag{5.21}$$

The filter output h_2^{ref} is the command compressor flow rate that provides the inner control loop.

Remark 5.5 It can be seen that (5.19) depends upon $\dot{\hat{\chi}}$. A real-time robust exact differentiator proposed in [18] can be used to estimate the time derivative of $\hat{\chi}$ in finite time. The differentiator has the following form

$$
\begin{aligned}
\dot{\zeta}_1 &= -\lambda_2 L^{\frac{1}{2}} |\zeta_1 - \hat{\chi}|^{\frac{1}{2}} \operatorname{sign}(\zeta_1 - \hat{\chi}) + \zeta_2, \\
\dot{\zeta}_2 &= -\alpha_2 L \operatorname{sign}(\zeta_1 - \hat{\chi}),
\end{aligned}
\tag{5.22}
$$

where ζ_1 and ζ_2 are the real-time estimations of $\hat{\chi}$ and $\dot{\hat{\chi}}$, respectively. The parameters of the differentiator $\lambda_2 = 1$, $\alpha_2 = 1.1$ are suggested and L is the only parameter needed to be tuned according to the condition $L \geq |\ddot{\hat{\chi}}|$.

Inner Loop

The inner loop will generate the compressor motor quadrature current u based on SMC. The sliding variable $s_2(t)$ is defined as follows:

$$
s_2(t) = h_2 - h_2^{\text{ref}}.
$$

Take the first time derivative of $s_2(t)$:

$$
\begin{aligned}
\dot{s}_2(t) &= c_{21} \left\{ -c_1 x_1 - \frac{c_2}{x_1} \left[\left(\frac{x_2}{c_3} \right)^{c_4} - 1 \right] h_2 + c_5 u \right\} - \dot{h}_2^{\text{ref}} \\
&= g_2 u + \varphi_2 (x_1, x_2) - \dot{h}_2^{\text{ref}},
\end{aligned}
\tag{5.23}
$$

where $g_2 = g_{20} + \Delta g_{20}$ and $\varphi_2 = \varphi_{20} + \Delta \varphi_{20}$ are smooth enough in the operation range, $\varphi_{20} > 0$ and g_{20} are known functions based on nominal parameters while $\Delta \varphi_{20}$ and Δg_{20} are parametric uncertainties. It is assumed that

$$
0 < \frac{|\Delta g_{20}|}{g_{20}} \leq b_2 < 1,
$$

and $\Delta \varphi_{20}$ is assumed to be bounded and with first derivative bounded [29].

Note that the sliding variable $s_2(t)$ has relative degree one with respect to the control input u, because u appears in the $\dot{s}_2(t)$ expression explicitly. Thus, the internal controller can be designed as follows:

$$
u = \frac{1}{g_{20}} \left(-k_1 \operatorname{sign}(s_2) - k_2 s_2 - \varphi_{20} \right),
\tag{5.24}
$$

for some positive constants k_1 and k_2. Applying (5.24) to (5.23), the sliding manifold can be rewritten as follows:

$$
\dot{s}_2 = -\gamma_2 k_1 \operatorname{sign}(s_2) - \gamma_2 k_2 s_2 + \phi_2,
$$

where $\gamma_2 = \left(1 + \frac{\Delta g_{20}}{g_{20}}\right)$ and $\phi_2 = \Delta\varphi_{20} - \frac{\Delta g_{20}}{g_{20}}\varphi_{20} - \dot{h}_2^{\text{ref}}$. Assume that the term ϕ_2 is bounded, i.e., $|\phi_2| \leq \Phi_2$, for some positive value Φ_2. The sufficient conditions for the finite time convergence to the sliding manifold $s_2(t) = \dot{s}_2(t) = 0$ are:

$$k_1 > \frac{\Phi_2}{1 - b_2}, \quad k_2 > 0.$$

In this study, the following boundary values are obtained as $\Phi_2 = 150$, $b_2 = 0.4$. Accordingly, the gains are tuned as $k_1 = 180$ and $k_2 = 25$.

Remark 5.6 It is clear to see that the control law (5.24) consists of a linear term $k_2 s_2$ and a switching term $k_1 \text{sign}(s_2)$. When the states trajectories are far from the surface, the linear term improves the behavior of the controller with a high rate of convergence. Conversely, the switching term dominates the the rate of convergence when the states trajectories are in the vicinity of the surface. Therefore, $k_2 > 0$ can be chosen according to the desired rate of convergence towards the sliding surface from the initial condition.

5.3.2   Closed-Loop Stability Analysis

Theorem 5.1 *Consider the system (5.3)–(5.7) in closed loop with the controller (5.24), satisfying the conditions in Lemma 5.1 and a smooth reference profile h_2^{ref} in (5.21). Then, there exists finite time $T > 0$ such that*

$$\left|\lambda_{O_2} - \lambda_{O_2}^*\right| \leq \delta_1(\delta, \mu), \quad t \geq T,$$

where δ_1 is a positive function depending on δ and μ.

Proof The detailed proof can be followed from [24].   ∎

Remark 5.7 The reason for introducing the smoothing filter (5.21) is that, the stability analysis of the internal compressor flow rate loop requires a bounded first time derivative of h_2^{ref} [24]. As can be seen from (5.19) that, h_2^* has a discontinuous time derivative because of the term $\sigma(s_1)$, thus, a first order filter can be used to perform the required smoothing action.

5.4   Experimental Results

The ESO-based control proposed in Sect. 5.3 is implemented on an HIL test bench shown in Fig. 4.5, where the implementable feasibility of the algorithm is evaluated. It consists of a physical air-feed system, based on a commercial twin screw compressor and a real-time PEMFC emulator derived from the work of [17, 20].

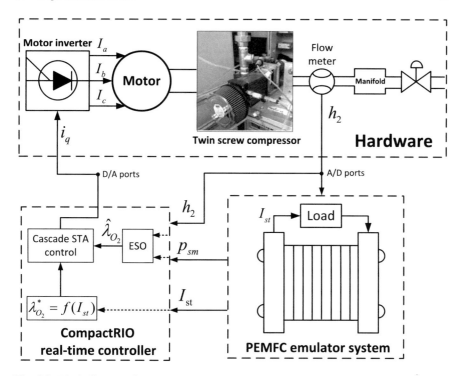

Fig. 5.3 Block diagram of the HIL test bench

The nominal values of the parameters for the HIL emulator are shown in Table 4.4. The PEMFC works under several safety constraints, the most important ones are the anode and cathode pressure difference (to be kept minimum) and stack temperature (to be regulated). These constraints r equire additional control in real fuel cells, which have been replicated in the emulator as well. An internal controller is implemented in the emulator to ensure that the anode pressure follows the cathode pressure at all times. The temperature is also controlled and its value can be set through software. The block diagram of this HIL test bench is shown in Fig. 5.3.

In order to test the proposed control's robustness against parametric uncertainties, the perturbations of the system parameters are given in Table 5.2, i.e., volumes, cathode inlet and outlet orifice, ambient temperature and stack temperature, motor constant and compressor inertia have been varied around their nominal values (see Table 4.4). During the tests, the load (stack current) is varied between 100 and 300 A as shown in Fig. 5.4. The corresponding compressor flow rate is shown in Fig. 5.5.

The results of the ESO-STA controller are compared with that of STA, as shown in Figs. 5.6, 5.7, 5.8, 5.9 and 5.10. The performance of the oxygen excess ratio by the proposed controller and STA is shown in Fig. 5.6. It is easy to see that the ESO is capable of reconstructing oxygen excess ratio in the presence of parameter uncertainties, which guarantees that the ESO-STA controller recovers the performance of

Table 5.2 Variations of system parameters

Symbol	Parameter	Value (%)
T_{st}	Temperature of the fuel cell, °C	+10
V_{ca}	Cathode volume, m³	+10
V_{sm}	Supply manifold volume, m³	−10
T_{atm}	Ambient temperature, °C	+10
$k_{ca,in}$	Cathode inlet orifice constant, kg/(Pa·s)	+5
$k_{ca,out}$	Cathode outlet orifice constant, kg/(Pa·s)	+5
k_t	Motor constant, N·m/A	−10
J_{cp}	Compressor inertia, kg·m²	+5

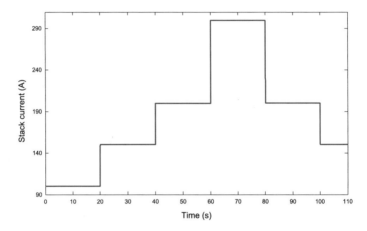

Fig. 5.4 Stack current under load variation

the state feedback STA controller when the gain β is chosen appropriately ($\beta = 1000$ is chosen). Little overshoot and fast convergence can also be observed from Fig. 5.6, i.e., at $t = 60$ s, sudden increase in load current (from 200 to 300 A) causes a sudden drop in the oxygen excess ratio (from 2.2 to 1.51). Conversely, at $t = 80$ s, sudden decrease in load current (from 300 to 200 A) causes a sudden increase in the oxygen excess ratio (from 2.2 to 2.91). It can be seen from Fig. 5.6a, b that the proposed controller has a good response and the error remains within an acceptable neighborhood of zero. In addition, the settling time of both upward load changes and downward load changes is less than 1.5 s for both controllers.

The performance of the Levant's differentiator [18] is shown in Fig. 5.7, where fast convergence and little overshoot can be observed. The compressor flow rate tracking its reference is shown in Fig. 5.8. The compressor flow rate reference is generated from the external regulation loop, which is based on the error between the desired value of oxygen excess ratio and its estimated value.

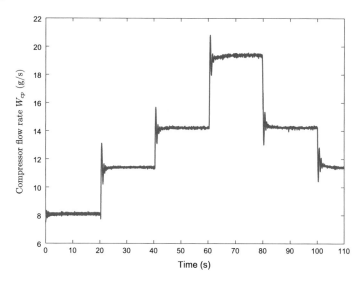

Fig. 5.5 Measurement of compressor flow rate under load variation

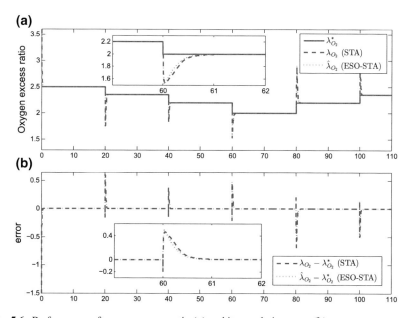

Fig. 5.6 Performance of oxygen excess ratio (**a**) and its regulation error (**b**)

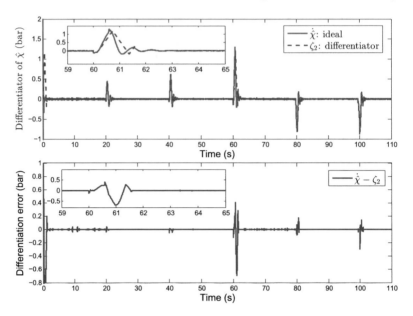

Fig. 5.7 Performance of the differentiator (5.22)

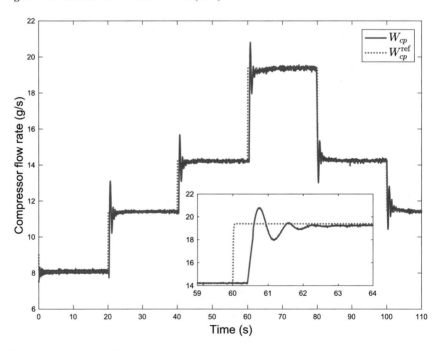

Fig. 5.8 Performance of the compressor flow rate

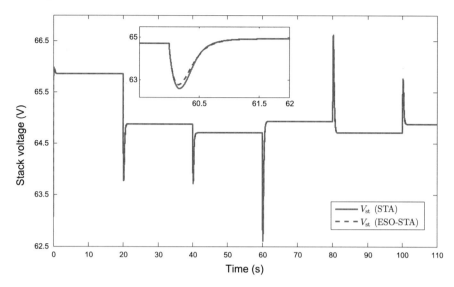

Fig. 5.9 Stack voltage performance

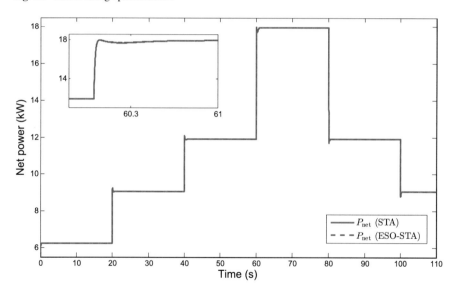

Fig. 5.10 Fuel cell net power

It should be pointed out that the output of the external regulation loop is compared with the experimental measurement. The similar behavior of the two controllers can be found, which again guarantees that the proposed ESO-STA controller can recover the performance of the state feedback STA controller. The associated stack voltages and net power of both controllers are shown in Figs. 5.9 and 5.10, respectively. The stack voltage varies between 62.5 and 66.5 V, while the net power of the fuel cell varies between 6.2 and 18 kW, according to the load current.

5.5 Conclusion

In this chapter, an ESO-based cascade controller for regulating the oxygen excess ratio of the PEMFC air feed system to its desired value, using sliding mode technique, has been presented. The control objective is to avoid oxygen starvation during sudden load changes, which is considered as a measurable disturbance in this work. The designed cascade controller consists of oxygen excess ratio tracking (outer) loop and compressor flow rate regulation (inner) loop.

Firstly, an ESO is employed to reconstruct the oxygen excess ratio with high accuracy and fast convergence. The outer control loop, which uses the estimated oxygen excess ratio, provides the compressor flow rate reference for the inner loop based on the STA. Secondly, a simple SMC law consists of a linear term and a switching term is designed for the inner loop, which ensures a fast response of the control scheme. Finally, the proposed ESO-based cascade control approach was successfully implemented on an HIL test bench. The experimental results have demonstrated the effectiveness, robustness and feasibility of the proposed controller. Furthermore, the obtained results are compared to state feedback STA control under the assumption that oxygen excess ratio is available, which show that the performance of the proposed ESO-based STA control can recover that of the state feedback control in the presence of parameter uncertainties. In the following chapter, nonlinear observer design for PEMFC systems using sliding mode technique will be investigated.

References

1. Arcak, M., Görgün, H., Pedersen, L.M., Varigonda, S.: A nonlinear observer design for fuel cell hydrogen estimation. IEEE Trans. Control Syst. Technol. **12**(1), 101–110 (2004)
2. Arce, A., del Real, A., Bordons, C., Ramirez, D.: Real-time implementation of a constrained MPC for efficient airflow control in a pem fuel cell. IEEE Trans. Ind. Electron. **57**(6), 1892–1905 (2010)
3. Bartolini, G., Pisano, A., Punta, E., Usai, E.: A survey of applications of second-order sliding mode control to mechanical systems. Int. J. Control **76**(9–10), 875–892 (2003)
4. Berning, T., Lu, D., Djilali, N.: Three-dimensional computational analysis of transport phenomena in a PEM fuel cell. J. Power Sources **106**(1), 284–294 (2002)
5. Carrette, L., Friedrich, K., Stimming, U.: Fuel cells-fundamentals and applications. Fuel Cells **1**(1), 5–39 (2001)

6. Damour, C., Benne, M., Lebreton, C., Deseure, J., Grondin-Perez, B.: Real-time implementation of a neural model-based self-tuning PID strategy for oxygen stoichiometry control in PEM fuel cell. Int. J. Hydrog. Energy **39**(24), 12819–12825 (2014)
7. Danzer, M., Wilhelm, J., Aschemann, H., Hofer, E.: Model-based control of cathode pressure and oxygen excess ratio of a PEM fuel cell system. J. Power Sources **176**(2), 515–522 (2008)
8. Garcia-Gabin, W., Dorado, F., Bordons, C.: Real-time implementation of a sliding mode controller for air supply on a PEM fuel cell. J. Process Control **20**(3), 325–336 (2010)
9. Gruber, J., Bordons, C., Oliva, A.: Nonlinear MPC for the airflow in a PEM fuel cell using a volterra series model. Control Eng. Pract. **20**(2), 205–217 (2012)
10. Han, J.: From PID to active disturbance rejection control. IEEE Trans. Ind. Electron. **56**(3), 900–906 (2009)
11. Hu, X., Murgovski, N., Johannesson, L.M., Egardt, B.: Optimal dimensioning and power management of a fuel cell/battery hybrid bus via convex programming. IEEE/ASME Trans. Mechatron. **20**(1), 457–468 (2015)
12. Hung, J.Y., Gao, W., Hung, J.C.: Variable structure control: a survey. IEEE Trans. Ind. Electron. **40**(1), 2–22 (1993)
13. Jang, M., Ciobotaru, M., Agelidis, V.G.: Design and implementation of digital control in a fuel cell system. IEEE Trans. Ind. Inform. **9**(2), 1158–1166 (2013)
14. Jemeï, S., Hissel, D., Pera, M.C., Kauffmann, J.M.: A new modeling approach of embedded fuel-cell power generators based on artificial neural network. IEEE Trans. Ind. Electron. **55**(1), 437–447 (2008)
15. Jung, J.H., Ahmed, S., Enjeti, P.: PEM fuel cell stack model development for real-time simulation applications. IEEE Trans. Ind. Electron. **58**(9), 4217–4231 (2011)
16. Kunusch, C., Puleston, P., Mayosky, M., Riera, J.: Sliding mode strategy for PEM fuel cells stacks breathing control using a super-twisting algorithm. IEEE Trans. Control Syst. Technol. **17**(1), 167–174 (2009)
17. Laghrouche, S., Liu, J., Ahmed, F.S., Harmouche, M., Wack, M.: Adaptive second-order sliding mode observer-based fault reconstruction for PEM fuel cell air-feed system. IEEE Trans. Control Syst. Technol. **23**(3), 1098–1109 (2015)
18. Levant, A.: Robust exact differentiation via sliding mode technique. Automatica **34**(3), 379–384 (1998)
19. Levant, A.: Higher-order sliding modes, differentiation and output-feedback control. Int. J. Control **76**(9–10), 924–941 (2003)
20. Matraji, I., Laghrouche, S., Jemei, S., Wack, M.: Robust control of the PEM fuel cell air-feed system via sub-optimal second order sliding mode. Appl. Energy **104**, 945–957 (2013)
21. Meidanshahi, V., Karimi, G.: Dynamic modeling, optimization and control of power density in a PEM fuel cell. Appl. Energy **93**, 98–105 (2012)
22. Muller, E.A., Stefanopoulou, A.G., Guzzella, L.: Optimal power control of hybrid fuel cell systems for an accelerated system warm-up. IEEE Trans. Control Syst. Technol. **15**(2), 290–305 (2007)
23. Pilloni, A., Pisano, A., Usai, E.: Observer-based air excess ratio control of a PEM fuel cell system via high-order sliding mode. IEEE Trans. Ind. Electron. **62**(8), 5236–5246 (2015)
24. Pisano, A., Davila, A., Fridman, L., Usai, E.: Cascade control of PM DC drives via second-order sliding-mode technique. IEEE Trans. Ind. Electron. **55**(11), 3846–3854 (2008)
25. Pukrushpan, J., Peng, H., Stefanopoulou, A.: Control-oriented modeling and analysis of fuel cell reactant flow for automotive fuel cell systems. ASME J. Dyn. Syst. Meas. Control **126**(1), 14–25 (2004)
26. Pukrushpan, J.T., Stefanopoulou, A.G., Peng, H.: Control of Fuel Cell Power Systems: Principles, Modeling, Analysis and Feedback Design. Springer Science & Business Media (2004)
27. Rakhtala, S.M., Noei, A.R., Ghaderi, R., Usai, E.: Design of finite-time high-order sliding mode state observer: a practical insight to PEM fuel cell system. J. Process. Control **24**(1), 203–224 (2014)
28. Ramos-Paja, C.A., Giral, R., Martinez-Salamero, L., Romano, J., Romero, A., Spagnuolo, G.: A pem fuel-cell model featuring oxygen-excess-ratio estimation and power-electronics interaction. IEEE Trans. Ind. Electron. **57**(6), 1914–1924 (2010)

29. Shtessel, Y., Taleb, M., Plestan, F.: A novel adaptive-gain supertwisting sliding mode controller: methodology and application. Automatica **48**(5), 759–769 (2012)
30. Suh, K.W., Stefanopoulou, A.G.: Performance limitations of air flow control in power-autonomous fuel cell systems. IEEE Trans. Control. Syst. Technol. **15**(3), 465–473 (2007)
31. Talole, S.E., Kolhe, J.P., Phadke, S.B.: Extended-state-observer-based control of flexible-joint system with experimental validation. IEEE Trans. Ind. Electron. **57**(4), 1411–1419 (2010)
32. Vahidi, A., Stefanopoulou, A., Peng, H.: Current management in a hybrid fuel cell power system: a model-predictive control approach. IEEE Trans. Control Syst. Technol. **14**(6), 1047–1057 (2006)
33. Vepa, R.: Adaptive state estimation of a PEM fuel cell. IEEE Trans. Energy Convers. **27**(2), 457–467 (2012)
34. Wilhelm, A.N., Surgenor, B.W., Pharoah, J.G.: Design and evaluation of a micro-fuel-cell-based power system for a mobile robot. IEEE/ASME Trans. Mechatron. **11**(4), 471–476 (2006)
35. Zhao, D., Gao, F., Bouquain, D., Dou, M., Miraoui, A.: Sliding-mode control of an ultrahigh-speed centrifugal compressor for the air management of fuel-cell systems for automotive applications. IEEE Trans. Veh. Technol. **63**(1), 51–61 (2014)
36. Zheng, Q., Chen, Z., Gao, Z.: A practical approach to disturbance decoupling control. Control. Eng. Pract. **17**(9), 1016–1025 (2009)

Chapter 6
Sliding Mode Observer of PEMFC Systems

As mentioned in the general introduction, it is not always possible to use sensors for measurements, either due to prohibitive costs of the sensing technology or because the quantity is not directly measurable. For precise control applications, state observer can be used for obtaining unavailable state values instead of sensors. A brief survey of existing methods in order to define the context of our work.

The problems addressed in this chapter are the design of SOSM observer and algebraic observer for PEMFC systems. For the SOSM observer, the goal is to estimate the hydrogen partial pressure in the anode channel of the PEMFC, using the measurements of stack voltage, stack current, anode pressure and anode inlet pressure. The proposed observer employs a nonlinear error injection term, where the error is obtained from the difference between the system voltage output obtained from an experimental validated nonlinear model and estimated voltage output obtained from the designed observer. The robustness of this observer against parametric uncertainties and load variations is studied, and the finite time convergence property is proved via Lyapunov analysis. For the algebraic observer, the goal is to estimate oxygen and nitrogen partial pressures in the fuel cell cathode side, using measurements of supply manifold pressure and compressor mass flow rate. As the proposed technique requires the time derivatives of the states, Lyapunov based adaptive HOSM differentiators are synthesized and implemented for estimating these derivatives without *a-priori* knowledge of the upper bounds of their higher order time derivatives.

6.1 Introduction

PEMFC is an electrochemical device that produces electricity from the chemical reaction between hydrogen and oxygen [16, 27]. The electrolyte membrane of a PEMFC has a special property that allows only positive protons to pass through while blocking electrons. They are under intensive development in the past few

© Springer Nature Switzerland AG 2020
J. Liu et al., *Sliding Mode Control Methodology in the Applications of Industrial Power Systems*, Studies in Systems, Decision and Control 249,
https://doi.org/10.1007/978-3-030-30655-7_6

years as they are regarded as potential alternative power sources in automotive applications due to their relatively small size, low temperature (40–180 °C), quick start up and easy manufacturing [41]. Recent developments in PEM and catalyst technology have greatly increased the power density of fuel cells, made them one of the most prominent technologies for future's automotive world [25].

Automotive fuel cell applications have more rigorous operating requirements than stationary applications [7], therefore these applications need precise control of performance, in order to guarantee the reliability, health, and safety of both, the fuel cell and the user. Along with control, health monitoring and safety systems are essential for the application of fuel cells in automotive systems. Both control and health-monitoring systems require precise measurements of different physical quantities in the fuel cell. However, it is not always possible to use sensors for measurements, either due to prohibitive costs of the sensing technology or because the quantity is not directly measurable. In both these cases, state observers serve as a replacement for physical sensors, for obtaining the unavailable quantities.

In PEMFC, hydrogen which is generated from the fuel processing system is fed into the anode side of the cell stack, while air is pumped into the cathode side through an air compressor. One hand, it is well known that the fuel cell stack life is reduced due to the so-called starvation phenomenon of the cell, because of insufficient supply of oxygen and hydrogen [49]. On the other hand, excessive supply is also undesirable due to the reduction of its efficiency. Pressure regulation remains one of the most challenging control problems. The main objective is to ensure minimum pressure difference between the anode and cathode side of the PEMFC, avoid the membrane from damage and increase the fuel cell stack life [5]. Several model based control approaches have been proposed for solving this problem, such as H_∞ robust control [31], feedback linearization plus pole placement [39], proportional integral (PI) plus static feed-forward controller [41] and static feedback controller [23]. To evaluate the performance of these controllers, a major obstacle is the absence of reliable measurements of hydrogen partial pressure, specially in the conditions of humidified gas streams inside the cell stack. However, it is not always possible to use sensors for measurements, either due to prohibitive costs of the sensing technology or because the quantity is not directly measurable. The sensors that do provide satisfactory performance usually suffer from slow response times, low accuracy, bulky and high cost [2]. Therefore, state observers serve as a replacement for physical sensors, are of great interest.

It is known that low oxygen partial pressure in the cathode reduces the generation capacity of fuel cell systems and affects the fuel cell stack life [35]. Therefore, the observation of oxygen partial pressure is essential for the feedback control and fault detection, in order to ensure the safety and longevity of the fuel cells [21, 44, 50]. However, most of the commercially available oxygen sensors do not operate properly in presence of humidified gas streams inside the fuel cell stack [2]. The sensors that do provide satisfactory performance are usually too big and costly. Therefore, observer designs for estimating the unmeasurable partial pressures are of great interest.

Many research endeavors have been focused during the recent years, on observation problems in fuel cell systems [8, 15]. Arcak et al. [2] developed an adaptive observer for hydrogen partial pressure estimation based on the fuel cell voltage. Görgün et al. [14] developed an algorithm for estimating partial pressures and the membrane water content in PEMFCs based on the resistive cell voltage drop. This algorithm has incorporated two adaptive observers for hydrogen and oxygen partial pressures, adapted from the work of Arcak et al. [2]. However, both of the above works lack robustness against the fuel cell voltage's measurement noise and the internal model relies upon unmeasurable values. Several kinds of KFs have been applied to the state estimation of fuel cell systems, i.e. classical KF [42], Unscented Kalman Filter (UKF) [38] and adaptive UKF [53]. These approaches are based on model linearization around pre-defined operating points of the system. Moreover, the calculation of Jacobian matrix of complex models like fuel cells are time consuming, and therefore difficult in real-time implementation [4, 22]. Ingimundarson et al. [19] proposed a model based estimation approach to hydrogen leak detection in PEMFCs. More recently, Linear Parameter Varying (LPV) observer was proposed by Lira et al. [32] for the application to fault detection in PEMFC systems, where the stack current was taken as the scheduling variable. Kunusch et al. [26] proposed an approach based on super-twisting algorithm, in order to estimate the hydrogen input flow at the anode of the stack and the water transport across the membrane. A Luenberger observer was employed by Thawornkuno and Panjapornpon [51] in order to estimate the membrane water content in PEMFCs. The main limitation of this method is that it can only converge to a neighborhood of the real system states in the presence of disturbances.

Due to the lack of a straightforward observer design method for a given nonlinear system, many observation methods are generally dependent upon state transformations, the structure of the system, the form of the nonlinearities and the boundedness of the system states [18]. Among the popular strategies, high gain observers [13] are usually employed to estimate the system states under the assumption that the nonlinearity vector is globally or locally Lipschitz. However, in practice, the Lipschitz constraint is not easy to obtain, which prevents the global convergence of the high gain observer. Although the circle-criterion observer design [3, 11] relaxes the requirement of Lipschitz constant, it remains limited to systems with positive-gradient nonlinearities.

SMOs have found wide application in the areas of parameter estimation [1], state estimation [47] and FDI [54, 55] in recent years. Their well-known advantages are robustness and insensitivity to external disturbance. HOSM Observers have better performance as compared to classical sliding mode based observers because their output is continuous and does not require low pass filters. The STA, which is an unique absolutely continuous sliding mode algorithm among the SOSM algorithms, therefore it does not suffer from the problem of chattering [28]. The main advantages of the STA are: it does not need the evaluation of the time derivative of the sliding variable; its continuous nature suppresses arbitrary disturbances with bounded time derivatives.

6.2 SMO Design for PEMFC Systems

6.2.1 Nonlinear Dynamic Model of PEMFC

Figure 6.1 shows a block diagram of a typical PEM fuel cell system [33]. It consists of four major subsystems: the hydrogen supply subsystem, the air feed subsystem, the humidifier subsystem and the cooling subsystem. As our study is related to the hydrogen supply subsystem, we will restrict the model for observer design specifically to the states of the hydrogen supply system and PEMFC anode. Certain assumptions have been imposed on the operating conditions, as follows [2, 43]:

A1. The stack temperature and humidity in the fuel cell cathode and anode are well controlled;
A2. The anode pressure is well controlled to follow the cathode pressure;
A3. The temperatures inside the anode and the cathode are equal to the stack temperature;
A4. The flow channel and cathode backing layer are lumped into one volume which assumes uniform conditions inside the anode channel;
A5. Vapor partial pressure in the stack is considered equal to the saturation pressure.

Remark 6.1 The first two assumptions are valid from practical point of view as the temperature, humidity and anode pressure are usually regulated externally in PEMFC applications. The third assumption is justified as the temperature dynamics of a PEMFC stack are slow [43]. Assumption 4 is employed because we are interested in estimating the hydrogen partial pressure in the exit, and not its distribution along the channel [2]. Assumption 5 means that if the gas humidity drops below 100%, liquid water will either evaporate into the cathode gas or it will accumulate in the cathode.

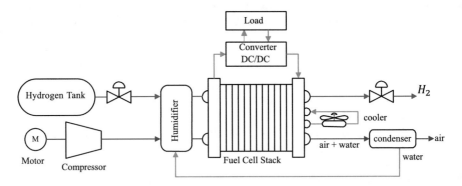

Fig. 6.1 Fuel cell system scheme

6.2.2 Gas Dynamics of Hydrogen Partial Pressure

Under these considerations, a lumped model is given as follows,

$$\dot{p}_{H_2} = \frac{RT_{fc}}{V_{an}M_{H_2}} \left(W_{H_2,in} - W_{H_2,out} - W_{H_2,react}\right), \tag{6.1}$$

where M_{H_2} is the molar masses of hydrogen $W_{H_2,in}$ is the mass flow rate of hydrogen gas entering the anode, $W_{H_2,out}$ is the mass flow rate of hydrogen gas leaving the anode, $W_{H_2,react}$ is the rate of hydrogen reacted, R is the universal gas constant, and V_{an} is the anode volume.

The inlet and outlet mass flow rates of hydrogen are given as follows,

$$W_{H_2,in} = \frac{1}{1+\omega_{an,in}} W_{an,in}, \tag{6.2}$$

$$W_{H_2,out} = \frac{1}{1+\omega_{an,out}} W_{an,out}, \tag{6.3}$$

where $\omega_{an,in}$ and $\omega_{an,in}$ are the anode inlet, outlet humidity ratios, respectively. $W_{an,in}$ and $W_{an,out}$ are anode inlet and outlet mass flow rates, respectively. Given that the purge of the anode gas is zero, they can be expressed as,

$$\begin{aligned} \omega_{an,in} &= \frac{M_v}{M_{H_2}} \frac{\phi_{an,in} p_{sat}(T_{fc})}{p_{an} - \phi_{an,in} p_{sat}(T_{fc})}, \\ W_{an,in} &= K_{an,in} \left(p_{an,in} - p_{an}\right), \\ W_{an,out} &= 0, \end{aligned} \tag{6.4}$$

where M_v is the molar masses of vapor, $\phi_{an,in}$ is the relative humidity on the anode, $p_{sat}(T_{fc})$ is the saturation pressure at fuel cell stack temperature. $K_{an,in}$ is the nozzle inlet flow constant for the anode side, $p_{an,in}$ is the anode inlet pressure, and p_{an} is the anode pressure.

The reacted mass flow rate is defined by,

$$W_{H_2,react} = M_{H_2} \frac{nI_{st}}{2F}, \tag{6.5}$$

where, F is Faraday constant, and I_{st} is the stack current, which is considered as a measurable disturbance variable.

6.2.3   SOSM Observer Design

In this Section, we are interested in the estimation of hydrogen partial pressure p_{H_2}, thus we will restrict the model for observer design specifically to the states of the PEMFC anode. We assume that the oxygen partial pressure p_{O_2} is available. This assumption is reasonable because we can simply employ a separate observer for the cathode channel, e.g. an adaptive numerical differentiation approach is proposed in Liu et al. [33] for estimating this value. Unlike partial pressures, the total pressures $(p_{an,in}, p_{an})$ are available for measurement. Therefore, an SOSM observer is developed for estimating the hydrogen partial pressure from the measurement of output stack voltage.

In view of Eqs. (4.1) and (4.2), it follows that

$$V_{fc} = n \left\{ E_0 + \frac{RT_{fc}}{2F} \left[\ln\left(p_{H_2}\right) + \frac{\ln\left(p_{O_2}\right)}{2} \right] - \nu(I_{st}) \right\}, \tag{6.6}$$

where $\nu(I_{st})$ represents the voltage drop as a function of current. The dynamics of p_{H_2} in (6.1) can be rewritten as,

$$\dot{p}_{H_2} = c_1 \left(p_{an,in} - p_{an}\right) - c_2 \zeta, \tag{6.7}$$

where $c_1 = \frac{R_{H_2} T_{fc}}{V_{an}} \frac{K_{an,in}}{1+\omega_{an,in}}$, $c_2 = \frac{R_{H_2} T_{fc}}{V_{an}} \frac{n M_{H_2}}{2F}$ and $\zeta := I_{st}$. Denote $\hat{p}_{H_2}$ as the estimate of hydrogen partial pressure. Since the voltage V_{fc} is a function of p_{H_2} and p_{O_2}, by substituting in $\hat{p}_{H_2}$ in (6.6), the voltage can be predicted as follows

$$\hat{V}_{fc} = n \left\{ E_0 + \frac{RT_{fc}}{2F} \left[\ln\left(\hat{p}_{H_2}\right) + \frac{\ln\left(p_{O_2}\right)}{2} \right] - \nu(\zeta) \right\}. \tag{6.8}$$

The output error term $\tilde{V}_{fc} = V_{fc} - \hat{V}_{fc}$ can be computed as

$$\tilde{V}_{fc} = \frac{nRT_{fc}}{2F} \left(\ln\left(p_{H_2}\right) - \ln\left(\hat{p}_{H_2}\right) \right). \tag{6.9}$$

Noting from the *Mean Value Theorem* that,

$$\ln\left(p_{H_2}\right) - \ln\left(\hat{p}_{H_2}\right) = \kappa \left(p_{H_2} - \hat{p}_{H_2}\right), \tag{6.10}$$

for some $\kappa \geq 1$. Substituting (6.10) into (6.8), we have,

$$\tilde{V}_{fc} = c_3 \left(p_{H_2} - \hat{p}_{H_2}\right), \tag{6.11}$$

where $c_3 = \frac{n\kappa RT_{fc}}{2F}$. The observer for estimating the hydrogen partial pressure $\hat{p}_{H_2}$ can be designed by adding the correction error term from the measurement of stack voltage,

$$\dot{\hat{p}}_{H_2} = c_1 \left(p_{an,in} - p_{an} \right) + \mu(\tilde{V}_{fc}), \tag{6.12}$$

where $\mu(s)$ is the output error-injection term generated from STA

$$\mu(s) = \lambda |s|^{\frac{1}{2}} \mathrm{sign}(s) + \alpha \int_0^t \mathrm{sign}(s) d\tau, \tag{6.13}$$

λ and α are some positive constants.

The observation error is defined as $e := p_{H_2} - \hat{p}_{H_2}$. Subtracting (6.12) from(6.7), the error dynamical equation is described by

$$\dot{e} = -\mu(c_3 e) - c_2 \zeta. \tag{6.14}$$

Theorem 6.1 *Consider the system (6.1), where the voltage V_{fc}, current I_{st}, temperature T_{fc}, partial oxygen pressure p_{O_2}, total anode pressure p_{an} and total anode inlet pressure $p_{an,in}$ are available. Then, the SOSM observer (6.12) and (6.13) guarantees the trajectories of the error system (6.14) to zero in finite time, given that λ and α are high enough.*

Proof The detailed proof can be followed from Moreno and Osorio [36] directly. ∎

Robustness Analysis

In this part, we will show the robustness of the proposed SOSM observer against model uncertainties and measurement disturbances. First, bounded disturbances will be introduced in the right-hand sides of (1) and (8), respectively. Thus, we have

$$\begin{aligned} \dot{p}_{H_2} &= c_1 \left(p_{an,in} - p_{an} \right) - c_2 \zeta + \rho(t), \\ V_{fc} &= V_{fc_0} + \delta(t), \end{aligned} \tag{6.15}$$

where V_{fc} denotes the nominal value in (6), $\delta(t)$ incorporates model inaccuracies and other disturbances acting on the plant, and $\epsilon(t)$ represents the measurement noise of stack voltage.

In view of (6.15) and (6.12), the error dynamical equation is described by

$$\dot{e} = -\mu \left(c_3 e + \epsilon(t) \right) + \rho(t), \tag{6.16}$$

where $\rho(t) = \delta(t) - c_2 \zeta$, the gains λ and α in the STA algorithm $\mu(\cdot)$ are formulated as

$$\lambda = \lambda_0 \sqrt{L}, \quad \alpha = \alpha_0 L, \tag{6.17}$$

where λ_0 and α_0 are positive constants and L is the only parameter needs to be tuned.

Theorem 6.2 *Assume that the conditions in Theorem 6.1 hold and the disturbances* $\rho(t)$ *and measurement noise* $\epsilon(t)$ *satisfy the following conditions*

$$|\dot{\rho}(t)| \leq \chi, \quad |\epsilon(t)| \leq \sigma. \tag{6.18}$$

Then, the following inequality is established in finite time,

$$\|\zeta\|_2 \leq \frac{\Omega_1(\sigma)}{\lambda_{min}(Q_1) - 4\|q_2\|_2 - 2\|q_3\|_2 \frac{\chi}{L}}, \tag{6.19}$$

if

$$L > \frac{2\chi\|q_3\|_2}{\lambda_{min}(Q_1) - \bar{\kappa} - 4\|q_2\|_2}, \tag{6.20}$$

where

$$\begin{aligned}
\Omega_1(\sigma) &= \|q_1\|_2(2\sigma)^{\frac{1}{2}}, & Q_1 &= \bar{\lambda}\begin{bmatrix} \bar{\lambda}^2 + 2\bar{\alpha} & -\bar{\lambda} \\ -\bar{\lambda} & 1 \end{bmatrix}, \\
q_1 &= [(\bar{\lambda}^2 + 4\bar{\alpha}), -\bar{\lambda}^2], & q_2 &= [-\bar{\lambda}\bar{\alpha}, 2\bar{\alpha}], \\
q_3 &= [-\bar{\lambda}, 2], & \bar{\lambda} &= c_3\lambda_0, & \bar{\alpha} &= c_3\alpha_0.
\end{aligned} \tag{6.21}$$

Proof Denote $e_1 := c_3 e$. The system (6.16) can be rewritten as

$$\begin{aligned}
\dot{e}_1 &= -\bar{\lambda}|e_1 + \epsilon|^{\frac{1}{2}}\operatorname{sign}(e_1 + \epsilon) + e_2, \\
\dot{e}_2 &= -\bar{\alpha}\operatorname{sign}(e_1 + \epsilon) + \dot{\bar{\rho}}(t),
\end{aligned} \tag{6.22}$$

where $e_1 = c_3 e$, $\bar{\lambda} = c_3\lambda$, $\bar{\alpha} = c_3\alpha$ and $\dot{\bar{\rho}}(t) = c_3\dot{\rho}(t)$.

A new state vector is introduced to represent the system (22) in a more convenient form for Lyapunov analysis.

$$\zeta = \begin{bmatrix} \zeta_1 \\ \zeta_2 \end{bmatrix} = \begin{bmatrix} L^{\frac{1}{2}}|e_1|^{\frac{1}{2}}\operatorname{sign}(e_1) \\ e_2 \end{bmatrix}. \tag{6.23}$$

Denote

$$\begin{aligned}
\Delta_1 &= |e_1|^{\frac{1}{2}}\operatorname{sign}(e_1) - |e_1 + \epsilon|^{\frac{1}{2}}\operatorname{sign}(e_1 + \epsilon), \\
\Delta_2 &= \operatorname{sign}(e_1) - \operatorname{sign}(e_1 + \epsilon).
\end{aligned} \tag{6.24}$$

In view of (6.18), it follows that

$$|\Delta_1| \leq \sqrt{2}\sigma^{\frac{1}{2}}, \quad |\Delta_2| \leq 2. \tag{6.25}$$

Then, the system (6.22) can be rewritten as follow

$$\dot{\zeta} = \frac{L}{2|\zeta_1|} \begin{bmatrix} -\bar{\lambda} & 1 \\ -2\bar{\alpha} & 0 \end{bmatrix} \zeta + \begin{bmatrix} \frac{\bar{\lambda}L^{\frac{3}{2}}}{2|\zeta_1|}\Delta_1 \\ \bar{\alpha}L\Delta_2 + \dot{\rho}(t) \end{bmatrix}. \tag{6.26}$$

The following Lyapunov function candidate is introduced for the system (6.26)

$$V(\zeta) = \zeta^T P \zeta, \quad P = \frac{1}{2} \begin{bmatrix} \bar{\lambda}^2 + 4\bar{\alpha} & -\bar{\lambda} \\ -\bar{\lambda} & 2 \end{bmatrix}, \tag{6.27}$$

where the matrix P is symmetric positive definite due to the fact that its leading principle minors are all positive. Taking the derivative of (6.27) yields

$$\dot{V} = -\frac{L}{2|\zeta_1|}\zeta^T Q_1 \zeta + \frac{L^{3/2}\Delta_1}{2|\zeta_1|}q_1\zeta + L\Delta_2 q_2\zeta + c_3 q_3\dot{\rho}(t)\zeta, \tag{6.28}$$

where

$$Q_1 = \bar{\lambda}\begin{bmatrix} \bar{\lambda}^2 + 2\bar{\alpha} & -\bar{\lambda} \\ -\bar{\lambda} & 1 \end{bmatrix}, \quad q_1 = [(\bar{\lambda}^2 + 4\bar{\alpha}), -\bar{\lambda}^2], \tag{6.29}$$

$$q_2 = [-\bar{\lambda}\bar{\alpha}, 2\bar{\alpha}], \quad q_3 = [-\bar{\lambda}, 2].$$

It is easy to verify that Q_1 is a positive definite matrix. Since $\lambda_{\min}(P)\|\zeta\|_2^2 \leq V \leq \lambda_{\max}(P)\|\zeta\|_2^2$, where $\|\cdot\|_2$ denotes the Euclidean norm of a vector or the spectral norm of a matrix. Then, Eq. (6.28) can be rewritten as

$$\dot{V} \leq -\frac{L}{2|\zeta_1|}\lambda_{min}(Q_1)\|\zeta\|_2^2 + \frac{L}{2|\zeta_1|}\Omega_1(\sigma)\|\zeta\|_2 \tag{6.30}$$
$$+ 2L\|q_2\|_2\|\zeta\|_2 + \chi\|q_3\|_2\|\zeta\|_2,$$

where $\Omega_1(\sigma) = \|q_1\|_2(2\sigma)^{\frac{1}{2}}$.

Suppose that there exists a positive constant $\bar{\kappa}$ such that

$$\Omega_1(\sigma) < \bar{\kappa}\|\zeta\|_2. \tag{6.31}$$

Therefore, it follows

$$\dot{V} \leq -\frac{L}{2|\zeta_1|}(\lambda_{\min}(Q_1) - \bar{\kappa})\|\zeta\|_2^2 + 2L\|q_2\|_2\|\zeta\|_2 + \chi\|q_3\|_2\|\zeta\|_2 \tag{6.32}$$
$$\leq -\gamma\|\zeta\|_2,$$

where $\gamma = \frac{L}{2}(\lambda_{\min}(Q_1) - \bar{\kappa}) - 2L\|q_2\|_2 - \chi\|q_3\|_2$. Under the condition (6.20), we have

$$\dot{V} \leq -\frac{\gamma}{\sqrt{\lambda_{\max}(P)}}V^{\frac{1}{2}}. \tag{6.33}$$

By the comparison principle [24], the differential inequality (6.32) is finite time convergent if

$$\|\zeta\|_2 > \frac{L\Omega_1(\sigma)}{L\lambda_{\min}(Q_1) - 4L\|q_2\|_2 - 2\chi\|q_3\|_2}. \tag{6.34}$$

Therefore, we get

$$\|\zeta\|_2 \leq \frac{L\Omega_1(\sigma)}{L\lambda_{\min}(Q_1) - 4L\|q_2\|_2 - 2\chi\|q_3\|_2}. \tag{6.35}$$

This completes the proof.   ∎

6.3 Algebraical Observer Design of PEMFC Systems

The main content in this chapter is the design of an algebraic observer for the PEMFC system in automotive applications. This observer is designed for observing the partial pressures of oxygen and nitrogen in the cathode of the PEMFC. The necessity of this type of observation arises typically in fuel-cell powered vehicles where air is used as oxygen source in the cathode of the PEMFC. As air is composed primarily of nitrogen and oxygen, the partial pressure of each gas needs to be estimated in order to ensure the existence of sufficient quantity of oxygen in the cathode for the chemical reaction. Insufficient oxygen quantity can result in low power output and also lead to permanent physical damage to the PEMFC [2]. As partial pressures cannot be measured from conventional sensors, their observation is essential for the control and fault detection, in order to ensure the safety and longevity of the PEMFC [21, 45, 50].

The motivation behind this work is that algebraic observers [10, 17] are precise and easily implementable in automotive embedded systems. We first demonstrate that the states of the PEMFC air-feed system is algebraically observable, i.e. they can be presented in terms of a static diffeomorphism [20] involving the system outputs (compressor flow rate and supply manifold pressure) and their time derivatives, respectively. As the algebraical observer requires the time derivatives of the output variables, its performance requires robust and exact differentiators [10]. The existing robust differentiators, such as Levant [29], require *a priori* knowledge of the upper bound of a higher order time derivative, i.e. the Lipschitz constant. In many practical cases, this boundary can not be easily obtained and the results are possibly very conservative. To overcome this problem, we propose new Lyapunov-based adaptive first and SOSM differentiators for practical implementation of our algebraic observation scheme. The adaptive gains adjust dynamically, therefore the Lipschitz constant is not required during the design.

The performance of the proposed observer is evaluated by implementing on the instrumented HIL test bench described in Chap. 4. In our experimental study, the main emphasis has been maintained on the robustness of the proposed observer against measurement noise and parameter variations. The use of PEMFC emulation system permits to conduct experiments on fuel cell auxiliary systems in real-time, while avoiding the risk of accidents (during worst case parametric variations) and cutting the consumption of expensive chemical reagents during fuel cell experiments that are not linked with fuel cell technology itself [12, 40, 46].

6.3.1 Algebraic Observer Design Using Sliding Mode Differentiators

Algebraic observers are ideal for implementation in real-time embedded systems because of their low computational requirements. The exact definition of algebraic observability is given in Definition 6.1. Our objective is to design an algebraical observer for the oxygen and nitrogen partial pressures in the PEMFC air-feed system from the available measurements of supply manifold pressure and compressor flow rate.

Let us briefly recall here that these observers are applicable to systems whose states can be expressed in terms of input and output variables and their time derivatives up to some finite degrees. Further details can be found in Diop et al. [10] and Ljung and Glad [34]. In this section, we will first demonstrate the algebraic observability of PEMFC air-feed system. Then, we will present Lyapunov-based adaptive HOSM differentiators for the implementation of the algebraical observer. Finally, using these differentiators, we will present the algebraical observer for this system.

6.3.1.1 Algebraic Observability

Definition 6.1 Consider the nonlinear system described by the following dynamic equations,

$$\begin{aligned} \dot{x}(t) &= f(x(t), u(t)), \\ y(t) &= h(x(t)), \end{aligned} \tag{6.36}$$

where $f(\cdot, \cdot) \in \mathbb{R}^n$ and $h(\cdot) \in \mathbb{R}^p$ are assumed to be continuously differentiable. $x(t) \in \mathbb{R}^n$ represents the system state vector, $u(t) \in \mathbb{R}^m$ is the control input vector and $y(t) \in \mathbb{R}^p$ is the output.

The system in (6.36) is said to be algebraically observable if there exist two positive integers μ and ν such that

$$x(t) = \phi\left(y, \dot{y}, \ddot{y}, \cdots, y^{(\mu)}, u, \dot{u}, \ddot{u}, \cdots, u^{(\nu)}\right), \tag{6.37}$$

where $\phi(\cdot) \in \mathbb{R}^n$ is a differentiable vector valued nonlinearity of the inputs, the outputs and their time derivatives [17].

Let us consider the model of PEMFC air-feed system (4.29). Define the cathode pressure p_{ca} as a new variable X, i.e. $X = x_1 + x_2 + c_2$. It follows from the last equation of (4.29) that

$$X = y_1 - \frac{y_2}{c_{16}} + \frac{\dot{y}_1}{c_{14}c_{16}\left\{1 + c_{15}\left[\left(\frac{y_1}{c_{11}}\right)^{c_{12}} - 1\right]\right\}} = \varphi_1(y_1, \dot{y}_1, y_2). \qquad (6.38)$$

The time derivative of X is given by

$$\begin{aligned}
\dot{X} = \dot{y}_1 &- \frac{\dot{y}_2}{c_{16}} + \frac{\ddot{y}_1}{c_{14}c_{16}\left\{1 + c_{15}\left[\left(\frac{y_1}{c_{11}}\right)^{c_{12}} - 1\right]\right\}} \\
&- \frac{c_{12}c_{15}y_1^{(c_{12}-1)}(\dot{y}_1)^2}{c_{11}^{c_{12}}c_{14}c_{16}\left\{1 + c_{15}\left[\left(\frac{y_1}{c_{11}}\right)^{c_{12}} - 1\right]\right\}^2} \\
&= \varphi_2(y_1, \dot{y}_1, \ddot{y}_1, \dot{y}_2).
\end{aligned} \qquad (6.39)$$

Then, in view of (6.38) and (6.39), the system states can be rewritten as

$$\begin{aligned}
x_1 &= \frac{1}{c_4 - c_5}\left[\frac{c_3(\varphi_1 - c_2)W_{ca,out}}{(c_1 + c_8)(y_1 - \varphi_1) - \varphi_2 - c_7\xi} + c_2c_5 - c_6 - c_5\varphi_1\right] \\
&= \varphi_3(y_1, \dot{y}_1, \ddot{y}_1, y_2, \dot{y}_2), \\
x_2 &= X - x_1 - c_2 = \varphi_4(y_1, \dot{y}_1, \ddot{y}_1, y_2, \dot{y}_2), \\
x_3 &= \frac{y_2}{c_{17}} = \varphi_5(y_2), \\
x_4 &= y_1 = \varphi_6(y_1).
\end{aligned} \qquad (6.40)$$

It can be seen from (6.40) that all states have been expressed as functions of the system outputs and a finite number of their time derivatives, i.e. $\dot{y}_1$, $\ddot{y}_1$ and $\dot{y}_2$. Therefore, according to the definition of algebraic observability 6.1, system (4.29) is algebraically observable.

6.3.1.2   Adaptive Sliding Mode Differentiators

Although system (6.40) is algebraically observable, it is necessary to estimate the time derivatives $\dot{y}_1$, $\ddot{y}_1$ and $\dot{y}_2$ accurately. In this work, $\dot{y}_1$ and $\ddot{y}_1$ are estimated through an adaptive SOSM differentiator and $\dot{y}_2$ is estimated through an adaptive first order differentiator. As discussed in the introduction, gain adaptation have been designed to overcome the requirement of knowledge of upper bounds on higher order time derivatives. The structure of both these differentiators is presented in the following subsections.

Adaptive second order differentiator for $y_1(t)$: Suppose that the output $y_1(t)$ is a smooth function, consider the following third order system [30]

$$
\begin{aligned}
\dot{z}_0 &= -\bar{\lambda}_0(t)\,|z_0 - y_1(t)|^{\frac{2}{3}}\,\text{sign}(z_0 - y_1(t)) + z_1, \\
\dot{z}_1 &= -\bar{\lambda}_1(t)\,|z_0 - y_1(t)|^{\frac{1}{3}}\,\text{sign}(z_0 - y_1(t)) + z_2, \qquad (6.41) \\
\dot{z}_2 &= -\bar{\lambda}_2(t)\text{sign}(z_0 - y_1(t)).
\end{aligned}
$$

Denote $\sigma_0 = z_0 - y_1(t)$, $\sigma_1 = z_1 - \dot{y}_1(t)$ and $\sigma_2 = z_2 - \ddot{y}_1(t)$, and then system (6.41) can be rewritten as

$$
\begin{aligned}
\dot{\sigma}_0 &= -\bar{\lambda}_0(t)\,|\sigma_0|^{\frac{2}{3}}\,\text{sign}(\sigma_0) + \sigma_1, \\
\dot{\sigma}_1 &= -\bar{\lambda}_1(t)\,|\sigma_0|^{\frac{1}{3}}\,\text{sign}(\sigma_0) + \sigma_2, \qquad (6.42) \\
\dot{\sigma}_2 &= -\bar{\lambda}_2(t)\text{sign}(\sigma_0) - \dddot{y}_1(t),
\end{aligned}
$$

where $\left|\dddot{y}_1(t)\right| \le L_3$ and L_3 is an *unknown* positive constant. The time varying gains $\bar{\lambda}_0(t)$, $\bar{\lambda}_1(t)$ and $\bar{\lambda}_2(t)$ are designed as follows

$$
\bar{\lambda}_0(t) = 2\bar{L}^{\frac{1}{3}}(t), \quad \bar{\lambda}_1(t) = 1.5\bar{L}^{\frac{2}{3}}(t), \quad \bar{\lambda}_2(t) = 1.1\bar{L}(t), \qquad (6.43)
$$

for some positive time varying scalar $\bar{L}(t)$ which is adapted according to

$$
\dot{\bar{L}}(t) = \begin{cases} \bar{k}, & \text{if } |\sigma_0| \ne 0, \\ 0, & \text{otherwise,} \end{cases} \qquad (6.44)
$$

where $\bar{k} > 0$ is an arbitrary positive design constant.

Proposition 6.1 *Consider the error system (6.42). Suppose that the gains $\bar{\lambda}_0(t)$, $\bar{\lambda}_1(t)$ and $\bar{\lambda}_2(t)$ satisfy (6.43) and (6.44). Then, the states of the error system (6.42) converge to zero in finite time, i.e., $\sigma_i = 0$, $i \in \{0, 1, 2\}$.*

Proof Consider the following Lyapunov function candidate for the error system (6.42)

$$
\bar{V}(\xi, \bar{L}(t)) = \bar{V}_0(\xi) + \frac{1}{4}\left(\bar{L}(t) - \bar{L}^*\right)^4, \qquad (6.45)
$$

where $\bar{V}_0(\xi)$ is based on the Lyapunov function proposed in Moreno [37]

$$
\bar{V}_0(\xi) = \xi^{\mathrm{T}}\Gamma\xi, \quad \Gamma = \begin{bmatrix} \gamma_1 & -\frac{1}{2}\gamma_{12} & 0 \\ -\frac{1}{2}\gamma_{12} & \gamma_2 & -\frac{1}{2}\gamma_{23} \\ 0 & -\frac{1}{2}\gamma_{23} & \gamma_3 \end{bmatrix}, \qquad (6.46)
$$

where $\xi^T = \left[|\sigma_0|^{\frac{2}{3}} \text{sign}(\sigma_0) \; \sigma_1 \; \sigma_2^2\right]$, coefficients $(\gamma_1, \gamma_{12}, \gamma_2, \gamma_{13}, \gamma_{23}, \gamma_3)$ are chosen such that $\bar{V}_0(\xi)$ is positive definite and radially unbounded. Since the adaptation law (6.44) makes the adaptive law $\bar{L}(t)$ bounded, it means that there exists a positive constant $\bar{L}^*$ such that $\bar{L}(t) < \bar{L}^*$ for $t \geq 0$.

The derivative of the Lyapunov function candidate (6.45) is given by

$$\dot{\bar{V}}(\xi, \bar{L}(t)) = \dot{\bar{V}}_0(\xi) - \bar{k} \left|\bar{L}(t) - \bar{L}^*\right|^3. \tag{6.47}$$

Under conditions given in Theorem 1 [37] and for some positive constant ϑ, we have $\dot{\bar{V}}_0(\xi) \leq -\vartheta \bar{V}_0^{\frac{3}{4}}(\xi)$. Apply Jensen's inequality $|x| + |y| \geq (|x|^q + |y|^q)^{\frac{1}{q}}$, $q = \frac{4}{3} > 1$, one gets

$$\dot{\bar{V}}(\xi, \bar{L}(t)) \leq -\vartheta \bar{V}_0^{\frac{3}{4}} - \bar{k}\left|\bar{L}(t) - \bar{L}^*\right|^3 \leq -\min\{\vartheta, 4^{\frac{3}{4}}\bar{k}\}\bar{V}^{\frac{3}{4}}. \tag{6.48}$$

According to the Theorem 4.2 in Bhat and Bernstein [6], it follows that ξ converges to zero in finite time, i.e. $\sigma_i = 0$, $i \in \{0, 1, 2\}$. This completes the proof. ∎

Adaptive first order differentiator for $y_2(t)$: Let us consider that the output $y_2(t)$ is a smooth function. An adaptive first-order sliding mode differentiator can be constructed for this signal as follows:

$$\begin{aligned}
\dot{\xi}_1 &= -\lambda(t) |e_1(t)|^{\frac{1}{2}} \text{sign}(e_1(t)) - k_\lambda(t)e_1(t) + \xi_2, \\
\dot{\xi}_2 &= -\alpha(t)\text{sign}(e_1(t)) - k_\alpha(t)e_1(t),
\end{aligned} \tag{6.49}$$

where $e_1(t) = \xi_1 - y_2(t)$. Denote $e_2(t) = \xi_2 - \dot{y}_2(t)$ and its dynamics can be rewritten as

$$\begin{aligned}
\dot{e}_1(t) &= -\lambda(t) |e_1(t)|^{\frac{1}{2}} \text{sign}(e_1(t)) - k_\lambda(t)e_1(t) + \xi_2, \\
\dot{e}_2(t) &= -\alpha(t)\text{sign}(e_1(t)) - k_\alpha(t)e_1(t) - \ddot{y}_2(t),
\end{aligned} \tag{6.50}$$

where $|\ddot{y}_2(t)| \leq L_2$, L_2 is an *unknown* positive constant and the time varying gains $\lambda(t), \alpha(t), k_\lambda(t)$ and $k_\alpha(t)$ are formulated as

$$\begin{aligned}
\lambda(t) &= \lambda_0\sqrt{L(t)}, & \alpha(t) &= \alpha_0 L(t), \\
k_\lambda(t) &= k_{\lambda_0} L(t), & k_\alpha(t) &= k_{\alpha_0} L^2(t),
\end{aligned} \tag{6.51}$$

for some positive constants $\lambda_0, \alpha_0, k_{\lambda_0}, k_{\alpha_0}$ and a positive, time-varying, scalar function $L(t)$ which is adapted according to

$$\dot{L}(t) = \begin{cases} k, & \text{if } |e_1(t)| \neq 0, \\ 0, & \text{otherwise,} \end{cases} \tag{6.52}$$

where k is an arbitrary positive constant.

Proposition 6.2 *Consider the error system (6.50). Suppose that the coefficients λ_0, α_0, k_{λ_0} and k_{α_0} in (6.51) satisfy the following condition:*

$$4\alpha_0 k_{\alpha_0} > 8k_{\lambda_0}^2 \alpha_0 + 9\lambda_0^2 k_{\lambda_0}^2. \tag{6.53}$$

Then, the states of the error system (6.50) converge to zero in finite time, i.e. $e_1(t) = \dot{e}_1(t) = 0$.

Proof A new state vector is introduced to represent the system (6.50) in a more convenient form for Lyapunov analysis.

$$\zeta = \begin{bmatrix} \zeta_1 \\ \zeta_2 \\ \zeta_3 \end{bmatrix} = \begin{bmatrix} L^{\frac{1}{2}}(t)|e_1(t)|^{\frac{1}{2}}\text{sign}(e_1(t)) \\ L(t)e_1(t) \\ e_2(t) \end{bmatrix}. \tag{6.54}$$

Thus, the system (6.50) can be rewritten as

$$\dot{\zeta} = \frac{L(t)}{|\zeta_1|}A_1\zeta + L(t)A_2\zeta + \begin{bmatrix} \frac{\dot{L}(t)}{2L(t)}\zeta_1 \\ \frac{\dot{L}(t)}{2L(t)}\zeta_2 \\ -\ddot{y}_2(t) \end{bmatrix}, \tag{6.55}$$

where

$$A_1 = \begin{bmatrix} -\frac{\lambda_0}{2} & 0 & \frac{1}{2} \\ 0 & -\lambda_0 & 0 \\ -\alpha_0 & 0 & 0 \end{bmatrix}, \quad A_2 = \begin{bmatrix} -\frac{k_{\lambda_0}}{2} & 0 & 0 \\ 0 & -k_{\lambda_0} & 1 \\ 0 & -k_{\alpha_0} & 0 \end{bmatrix}. \tag{6.56}$$

Then, the following Lyapunov function candidate is introduced for the system (6.55)

$$V(\zeta) = \zeta^T P \zeta, \quad P = \frac{1}{2}\begin{bmatrix} 4\alpha_0 + \lambda_0^2 & \lambda_0 k_{\lambda_0} & -\lambda_0 \\ \lambda_0 k_{\lambda_0} & k_{\lambda_0}^2 + 2k_{\alpha_0} & -k_{\lambda_0} \\ -\lambda_0 & -k_{\lambda_0} & 2 \end{bmatrix}, \tag{6.57}$$

where the matrix P is positive definite. Taking the derivative of (6.57) yields

$$\dot{V} = -L(t)\left(\frac{\zeta^T \Omega_1 \zeta}{|\zeta_1|} + \zeta^T \Omega_2 \zeta\right) - \ddot{y}_2(t)q_1\zeta + \frac{q_2\dot{L}(t)}{L(t)}P\zeta, \tag{6.58}$$

where

$$q_1 = \begin{bmatrix} -\lambda_0 & -k_{\lambda_0} & 2 \end{bmatrix}, \quad q_2 = \begin{bmatrix} \zeta_1 & \zeta_2 & 0 \end{bmatrix},$$

$$\Omega_1 = \frac{\lambda_0}{2} \begin{bmatrix} \lambda_0^2 + 2\alpha_0 & 0 & -\lambda_0 \\ 0 & 2k_{\alpha_0} + 5k_{\lambda_0}^2 & -3k_{\lambda_0} \\ -\lambda_0 & -3k_{\lambda_0} & 1 \end{bmatrix},$$

$$\Omega_2 = k_{\lambda_0} \begin{bmatrix} \alpha_0 + 2\lambda_0^2 & 0 & 0 \\ 0 & k_{\alpha_0} + k_{\lambda_0}^2 & -k_{\lambda_0} \\ 0 & -k_{\lambda_0} & 1 \end{bmatrix}. \tag{6.59}$$

It is easy to verify that Ω_1 and Ω_2 are positive definite matrices under the condition (6.53).

Since $\lambda_{\min}(P)\|\zeta\|^2 \leq V \leq \lambda_{\max}(P)\|\zeta\|^2$, Eq. (6.58) can be rewritten as

$$\dot{V} \leq -L(t)\frac{\lambda_{\min}(\Omega_1)}{\lambda_{\max}^{\frac{1}{2}}(P)}V^{\frac{1}{2}} - L(t)\frac{\lambda_{\min}(\Omega_2)}{\lambda_{\max}(P)}V \tag{6.60}$$

$$+ \frac{L_2\|q_1\|_2}{\lambda_{\min}^{\frac{1}{2}}(P)}V^{\frac{1}{2}} + \frac{\dot{L}(t)}{2L(t)}\zeta^{\mathsf{T}}Q\zeta,$$

where

$$Q = \begin{bmatrix} 4\alpha_0 + \lambda_0^2 + \lambda_0 k_{\lambda_0} + \frac{\lambda_0}{2} & 0 & 0 \\ 0 & \frac{\lambda_0 + k_{\lambda_0}}{2} & 0 \\ 0 & 0 & 2k_{\alpha_0}k_{\lambda_0}^2 + \lambda_0 k_{\lambda_0} + \frac{k_{\lambda_0}}{2} \end{bmatrix}. \tag{6.61}$$

In view of (6.61), (6.60) can be rewritten as

$$\dot{V} \leq -(L(t)\gamma_1 - \gamma_2)V^{\frac{1}{2}} - \left(L(t)\gamma_3 - \gamma_4\frac{\dot{L}(t)}{L(t)}\right)V, \tag{6.62}$$

where

$$\gamma_1 = \frac{\lambda_{\min}(\Omega_1)}{\lambda_{\max}^{\frac{1}{2}}(P)}, \quad \gamma_2 = \frac{L_2\|q_1\|_2}{\lambda_{\min}^{\frac{1}{2}}(P)},$$

$$\gamma_3 = \frac{\lambda_{\min}(\Omega_2)}{\lambda_{\max}(P)}, \quad \gamma_4 = \frac{\lambda_{\max}(Q)}{2\lambda_{\min}(P)}. \tag{6.63}$$

Given that $\dot{L}(t) \geq 0$, thus the terms $L(t)\gamma_1 - \gamma_2$ and $L(t)\gamma_3 - \frac{\dot{L}}{L(t)}\gamma_4$ are positive in finite time, it follows from (6.62) that

$$\dot{V} \leq -c_1 V^{\frac{1}{2}} - c_2 V, \tag{6.64}$$

where c_1 and c_2 are positive constants. By the comparison principle [24], it follows that ζ converge to zero in finite time, i.e. $e_1(t) = 0$ and $e_2(t) = 0$. Thus, Proposition 6.2 is proven. ∎

Propositions 6.1 and 6.2 imply that the states of the adaptive second order differentiator (6.41), (6.43), (6.44) and first order differentiator (6.49), (6.51), (6.52) converge to their real values in finite time, i.e. $z_1 \to \dot{y}_1(t)$, $z_2 \to \ddot{y}_1(t)$ and $\xi_2 \to \dot{y}_2(t)$, respectively.

The algebraical observer is formulated as the following theorem.

Theorem 6.3 *For any $u \in U$ such that $y_1(t)$ and $y_2(t)$ are continuously differentiable, the following dynamic system*

$$
\begin{aligned}
\hat{x}_1 &= \varphi_3 \left(y_1(t), z_1, z_2, y_2(t), \xi_2 \right), \\
\hat{x}_2 &= \varphi_4 \left(y_1(t), z_1, z_2, y_2(t), \xi_2 \right), \\
\hat{x}_3 &= \varphi_5 (y_2(t)), \\
\hat{x}_4 &= \varphi_6 (y_1(t)),
\end{aligned}
\tag{6.65}
$$

where ξ_2 and z_1, z_2 are the estimates of $\dot{y}_2(t)$, $\dot{y}_1(t)$ and $\ddot{y}_1(t)$ respectively, is a finite time observer for system (4.29).

Proof From the results of Propositions 6.1 and 6.2, it follows that there exists a finite time t_F after that ξ_2, z_1 and z_2 converge to $\dot{y}_2(t)$, $\dot{y}_1(t)$ and $\ddot{y}_1(t)$ exactly. Consequently, $\lim\limits_{t \geq t_F} \hat{x}_i = x_i$, $i = \{1, 2, 3, 4\}$. ∎

Remark 6.2 In practical implementation, ideal sliding mode is not achievable due to measurement noise and numerical approximation errors. Moreover, the adaptation laws in (6.44) and (6.52) will result in unbounded $L(t)$ and $\bar{L}(t)$. A feasible alternative to overcome this disadvantage is to modify the adaptive laws by dead zone technique [48, 52] as follows

$$
\dot{\bar{L}}(t) = \begin{cases} \bar{k}, & \text{if } |\sigma_0| \geq \bar{\epsilon}, \\ 0, & \text{otherwise}, \end{cases} \qquad \dot{L}(t) = \begin{cases} k, & \text{if } |e_1(t)| \geq \epsilon, \\ 0, & \text{otherwise}, \end{cases}
\tag{6.66}
$$

where ϵ and $\bar{\epsilon}$ are sufficient small positive constants. In this sense, the gains $\bar{L}(t)$ and $L(t)$ will be adapted dynamically according to (6.66) and σ_0 and $e_1(t)$ will be forced back into a real sliding mode regime in finite time.

6.4   Simulation Results of SMO for PEMFC Systems

A completely nonlinear model developed is used in our observer design, comprising of the following state vector:

$$
x = \begin{bmatrix} p_{O_2}, & p_{H_2}, & p_{N_2}, & p_{sm}, & m_{sm}, & p_{v,an}, & p_{v,ca}, & p_{rm}, & T_{st} \end{bmatrix}^{\mathrm{T}},
\tag{6.67}
$$

where p_{H_2} is hydrogen pressure in the anode, m_{sm} is mass of air in supply manifold, $p_{v,an}$ is vapor partial pressure in the anode, $p_{v,ca}$ is vapor partial pressure in the

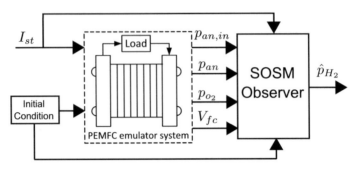

Fig. 6.2 Schematic diagram of the proposed SOSM observer

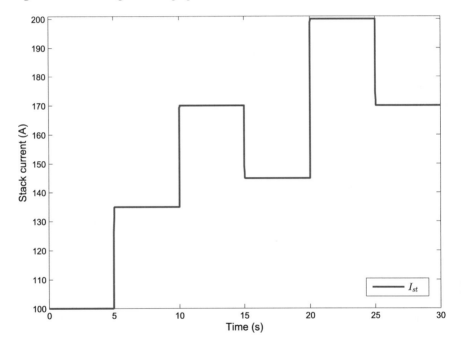

Fig. 6.3 Stack current under load variation

cathode, p_{rm} is pressure of return manifold and T_{st} is the stack temperature. The nominal values of the parameters used in the simulation are shown in Table 6.1.

The dynamic model is based on Pukrushpan et al. [43], with the added temperature model described by a lumped thermal model [9]:

$$\frac{dT_{st}}{dt} = \frac{\dot{Q}_{sou} - W_c C_{p_c} \left(T_{st} - T_{c,in} \right)}{m_{st} C_{p_{st}}},$$

$$\dot{Q}_{sou} = I_{st} \left(-\frac{T \Delta s}{4F} + v_{act} + I_{st} R_{ohm} \right),$$

(6.68)

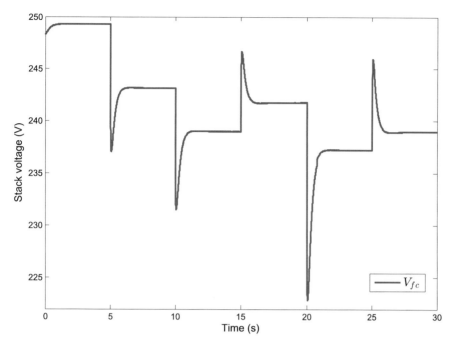

Fig. 6.4 Stack voltage response under load variations

where m_{st} is the heat mass of the stack, $C_{p_{st}}$ and C_{p_c} are the specific heat, W_c is the coolant flow rate considered as a control variable, $T_{c,in}$ is the coolant temperature at the stack inlet, v_{act} is the activation loss at catalyst layer and $\dot{Q}_{sou}$ is the internal energy source. The latter is calculated as a function of the stack current, temperature, electrical resistance of stack layers R_{ohm}, Faraday's number F and the entropy change Δs. The physical parameters were obtained through extensive experimentation.

In order to obtain the numerical solution, the initial conditions of the state vector (6.67) are given as:

$$
\begin{aligned}
p_{O_2}(0) &= 0.10\,\text{bar}, & p_{H_2}(0) &= 1.1\,\text{bar}, \\
p_{N_2}(0) &= 0.68\,\text{bar}, & p_{sm}(0) &= 1.48\,\text{bar}, \\
m_{sm}(0) &= 0.03\,\text{kg}, & p_{v,an}(0) &= 0.41\,\text{bar}, \\
p_{v,ca}(0) &= 0.63\,\text{bar}, & p_{rm}(0) &= 1.3\,\text{bar}, \quad T_{st}(0) = 340\,\text{K}.
\end{aligned}
\tag{6.69}
$$

In order to test the proposed observer's robustness against parametric uncertainties, the system parameters, i.e. volumes, anode inlet orifice, ambient temperature and stack temperature have been varied around their nominal values according to the worst-case percentages are given in Table 6.2. The schematic diagram of the proposed SOSM observer is shown in Fig. 6.2, where the observer gains are set as $\lambda_0 = 5$, $\alpha_0 = 10$ and $L = 25$, respectively. The initial errors of the states are set at

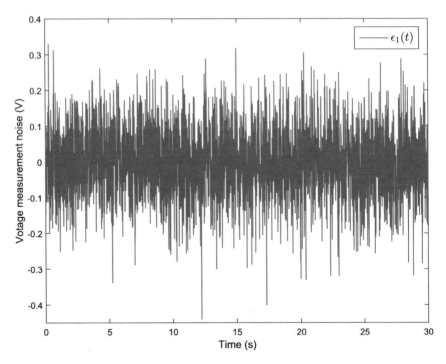

Fig. 6.5 Voltage measurement noise

Table 6.1 PEMFC system parameters

Symbol	Value	Symbol	Value
n	381	R	8.314 J/(mol·K)
R_a	286.9 J/(kg·K)	p_{atm}	1.01325 bar
T_{amb}	298.15 K	F	96485 C/mol
M_{H_2}	2 g/mol	M_v	18.02 g/mol
M_{O_2}	32 g/mol	M_a	28.96 g/mol
M_{N_2}	28 g/mol	R_{N_2}	296.8 J/(kg K)
R_{O_2}	259.8 J/(kg K)	R_{H_2}	4124.3 J/(kg K)
R_v	461.5 J/(kg K)	$p_{sat}(T_{fc} = 80\,°C)$	0.4707 bar
V_{an}	0.005 m^3	V_{ca}	0.01 m^3
$K_{an,in}$	2.1×10^{-3} kg/(Pa·s)	$K_{ca,in}$	3.629×10^{-6} kg/(Pa·s)

20% of maximum deviation from the fuel cell system. For simulation purposes, the initial values were chosen as $\hat{p}_{H_2}(0) = 0.9$bar.

During the tests, the load, i.e. the stack current shown in Fig. 6.3. was varied between 100 A and 200A. The stack voltage shown in Fig. 6.4 varies between 222 V

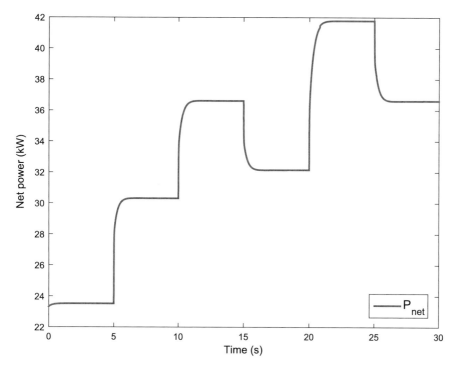

Fig. 6.6 Net power under load variations

Table 6.2 Parameter variations of the PEMFC system

Symbol	Parameter	Value (%)
T_{fc}	Temperature of the fuel cell, $°C$	+10
V_{an}	Cathode volume, m^3	+10
V_{sm}	Supply manifold volume, m^3	−10
T_{amb}	Ambient temperature, $°C$	+10
$K_{an,in}$	Cathode inlet orifice constant, kg/(Pa·s)	+5

and 248V. Additional noise was added to the measured signal $V_{fc} = V_{fc_0} + \epsilon_1(t)$, as shown in Fig. 6.5. The FC net power varies with the load current (Fig. 6.3), between 23.3 kW and 41.5 kW as shown in Fig. 6.6. The performance of oxygen excess ratio is given in Fig. 6.7. It can be easily found that the oxygen excess ratio overshoots at t = 15 s with respect to the step decrease of load current (Fig. 6.3). This causes an overshoot in the stack voltage (Fig. 6.4). Figure 6.8 shows that the state variable (p_{H_2}) are well estimated by the proposed observer under step load variations and measurement noise. Figure 6.9 shows the response of observation error based on the

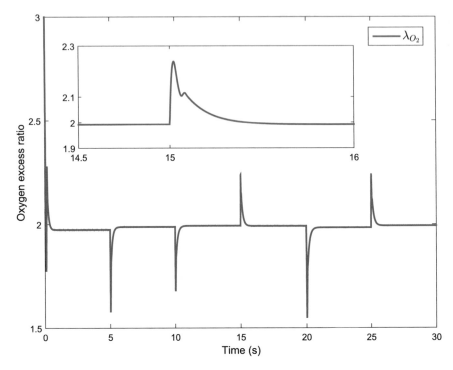

Fig. 6.7 Performance of oxygen excess ratio

measurements (p_{an}, $p_{an,in}$, p_{O_2}, I_{st} and V_{fc}). The initial errors of the states are set at 15% of maximum deviation from the fuel cell system. It is clear from the figures, the proposed observer is robust against model uncertainties and the effect of the noise is essentially imperceptible.

6.5   Conclusions

In this chapter, two nonlinear observers, i.e., An SOSM observer and an algebraic observer based on adaptive sliding mode differentiators have been designed for PEMFC system. Firstly, with the help of the SOSM observer, the hydrogen partial pressure was successfully observed from the measurements of anode inlet pressure ($p_{an,in}$), anode total pressure (p_{an}), oxygen partial pressure (p_{O_2}), load current (I_{st}) and stack output voltage (V_{fc}). It shows that the proposed observer is insensitive to model uncertainty and measurement noise. The effectiveness, robustness and feasibility of the SOSM observer under load variations, have been validated through simulation results.

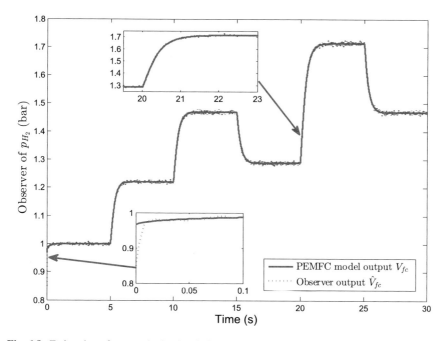

Fig. 6.8 Estimation of p_{H_2} under load variations and measurement noise

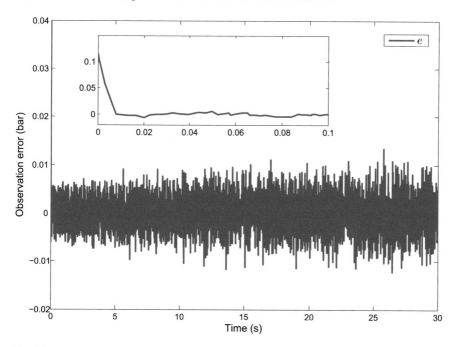

Fig. 6.9 Estimation error of hydrogen partial pressure

Secondly, the design of an algebraic observer has been presented. The system states, oxygen, nitrogen partial pressures, compressor angular speed and supply manifold pressure, are expressed in terms of input and output variables and a finite number of their time derivatives. These variables are estimated via adaptive differentiators without any knowledge of the upper bounds of their higher-order time derivatives. The oxygen and nitrogen partial pressures have been successfully observed from the measurements of stack current, supply manifold pressure and compressor flow rate. The effectiveness and feasibility of the proposed observer has been validated. In the next chapter, we will focus on the design of fault diagnosis approaches for PEMFC systems using sliding mode technique.

References

1. Adetola, V., Guay, M.: Finite-time parameter estimation in adaptive control of nonlinear systems. IEEE Trans. Autom. Control. **53**(3), 807–811 (2008)
2. Arcak, M., Görgün, H., Pedersen, L.M., Varigonda, S.: A nonlinear observer design for fuel cell hydrogen estimation. IEEE Trans. Control. Syst. Technol. **12**(1), 101–110 (2004)
3. Arcak, M., Kokotovic, P.: Observer-based control of systems with slope-restricted nonlinearities. IEEE Trans. Autom. Control. **46**(7), 1146–1150 (2001)
4. Arce, A., Alejandro, J., Bordons, C., Ramirez, D.R.: Real-time implementation of a constrained MPC for efficient airflow control in a PEM fuel cell. IEEE Trans. Ind. Electron. **57**(6), 1892–1905 (2009)
5. Barbir, F., Gorgun, H., Wang, X.: Relationship between pressure drop and cell resistance as a diagnostic tool for PEM fuel cells. J. Power Sources **141**(1), 96–101 (2005)
6. Bhat, S., Bernstein, D.: Finite-time stability of continuous autonomous systems. SIAM J. Control. Optim. **38**(3), 751–766 (2000)
7. Chen, F., Chu, H.S., Soong, C.Y., Yan, W.M.: Effective schemes to control the dynamic behavior of the water transport in the membrane of PEM fuel cell. J. Power Sources **140**(2), 243–249 (2005)
8. Chen, P.C.: Output-feedback voltage tracking control for input-constrained PEM fuel cell systems. Int. J. Hydrog. Energy **36**(22), 14608–14621 (2011)
9. Choe, S.Y., Ahn, J.W., Lee, J.G., Baek, S.H.: Dynamic simulator for a PEM fuel cell system with a PWM DC/DC converter. IEEE Trans. Energy Convers. **23**(2), 669–680 (2008)
10. Diop, S., Grizzle, J., Moraal, P., Stefanopoulou, A.: Interpolation and numerical differentiation for observer design. In: Proceedings of the American Control Conference, vol. 2, pp. 1329–1329. American Automatic Control Council (1994)
11. Fan, X., Arcak, M.: Observer design for systems with multivariable monotone nonlinearities. Syst. Control. Lett. **50**(4), 319–330 (2003)
12. Gao, F., Blunier, B., Simoes, M.G., Miraoui, A.: PEM fuel cell stack modeling for real-time emulation in hardware-in-the-loop applications. IEEE Trans. Energy Convers. **26**(1), 184–194 (2011)
13. Gauthier, J.P., Hammouri, H., Othman, S.: A simple observer for nonlinear systems applications to bioreactors. IEEE Trans. Autom. Control. **37**(6), 875–880 (1992)
14. Görgün, H., Arcak, M., Barbir, F.: An algorithm for estimation of membrane water content in PEM fuel cells. J. Power Sources **157**(1), 389–394 (2006)
15. Görgün, H., Arcak, M., Varigonda, S., Bortoff, S.A.: Observer designs for fuel processing reactors in fuel cell power systems. Int. J. Hydrog. Energy **30**(4), 447–457 (2005)
16. Grove, W., Egeland, O.: A small voltaic battery of great energy. Philos. Mag. **15**, 287–293 (1839)

17. Ibrir, S.: Online exact differentiation and notion of asymptotic algebraic observers. IEEE Trans. Autom. Control **48**(11), 2055–2060 (2003)
18. Ibrir, S.: Algebraic observer design for a class of uniformly-observable nonlinear systems: application to 2-link robotic manipulator. In: 2009 7th Asian Control Conference, pp. 390–395. IEEE (2009)
19. Ingimundarson, A., Stefanopoulou, A.G., McKay, D.A.: Model-based detection of hydrogen leaks in a fuel cell stack. IEEE Trans. Control. Syst. Technol **16**(5), 1004–1012 (2008)
20. Isidori, A.: Nonlinear control systems, vol. 1. Springer (1995)
21. Jemeï, S., Hissel, D., Pera, M.C., Kauffmann, J.M.: A new modeling approach of embedded fuel-cell power generators based on artificial neural network. IEEE Trans. Ind. Electron. **55**(1), 437–447 (2008)
22. Jung, J.H., Ahmed, S., Enjeti, P.: PEM fuel cell stack model development for real-time simulation applications. IEEE Trans. Ind. Electron. **58**(9), 4217–4231 (2011)
23. Karnik, A.Y., Sun, J., Stefanopoulou, A.G., Buckland, J.H.: Humidity and pressure regulation in a PEM fuel cell using a gain-scheduled static feedback controller. IEEE Trans. Control. Syst. Technol. **17**(2), 283–297 (2009)
24. Khalil, H.K.: Nonlinear Systems. Prentice Hall (2001)
25. Kirubakaran, A., Jain, S., Nema, R.K.: A review on fuel cell technologies and power electronic interface. Renew. Sustain. Energy Rev. **13**(9), 2430–2440 (2009)
26. Kunusch, C., Moreno, J., Angulo, M.: Identification and observation in the anode line of PEM fuel cell stacks. In: IEEE 52nd Annual Conference on Decision and Control (CDC), pp. 1665–1670. IEEE (2013)
27. Larminie, J., Dicks, A., McDonald, M.S.: Fuel Cell Systems Explained, vol. 2. Wiley, Chichester (2003)
28. Levant, A.: Sliding order and sliding accuracy in sliding mode control. Int. J. Control **58**(6), 1247–1263 (1993)
29. Levant, A.: Robust exact differentiation via sliding mode technique. Automatica **34**(3), 379–384 (1998)
30. Levant, A.: Higher-order sliding modes, differentiation and output-feedback control. Int. J. Control **76**(9–10), 924–941 (2003)
31. Li, Q., Chen, W., Wang, Y., Jia, J., Han, M.: Nonlinear robust control of proton exchange membrane fuel cell by state feedback exact linearization. J. Power Sources **194**(1), 338–348 (2009)
32. Lira, S.D., Puig, V., Quevedo, J., Husar, A.: LPV observer design for PEM fuel cell system: Application to fault detection. J. Power Sources **196**(9), 4298–4305 (2011)
33. Liu, J., Laghrouche, S., Ahmed, F.S., Wack, M.: PEM fuel cell air-feed system observer design for automotive applications: an adaptive numerical differentiation approach. Int. J. Hydrog. Energy **39**(30), 210–17 (2014)
34. Ljung, L., Glad, T.: On global identifiability for arbitrary model parametrizations. Automatica **30**(2), 265–276 (1994)
35. Matraji, I., Laghrouche, S., Wack, M.: Pressure control in a PEM fuel cell via second order sliding mode. Int. J. Hydrog. Energy **37**(21), 16104–16116 (2012)
36. Moreno, J., Osorio, M.: A Lyapunov approach to second-order sliding mode controllers and observers. In: 47th IEEE Conference on Decision and Control (CDC), pp. 2856–2861. IEEE (2008)
37. Moreno, J.A.: Lyapunov function for Levant's second order differentiator. In: IEEE 51st Annual Conference on Decision and Control (CDC), pp. 6448–6453. IEEE (2012)
38. Murshed, A., Huang, B., Nandakumar, K.: Estimation and control of solid oxide fuel cell system. Comput. Chem. Eng. **34**(1), 96–111 (2010)
39. Na, W.K., Gou, B.: Feedback-linearization-based nonlinear control for pem fuel cells. IEEE Trans. Energy Convers. **23**(1), 179–190 (2008)
40. Paja, C.A.R., Nevado, A.R., Castillón, R.G., Martinez-Salamero, L., Saenz, C.I.S.: Switching and linear power stages evaluation for PEM fuel cell emulation. Int. J. Circuit Theory Appl. **39**(5), 475–499 (2011)

41. Pukrushpan, J., Peng, H., Stefanopoulou, A.: Control-oriented modeling and analysis for automotive fuel cellsystems. ASME J. Dyn. Syst. Meas. Control. **126**(1), 14–25 (2004)
42. Pukrushpan, J., Stefanopoulou, A., Peng, H.: Control of fuel cell breathing. IEEE Control Syst. **24**(2), 30–46 (2004)
43. Pukrushpan, J., Stefanopoulou, A., Peng, H.: Control of fuel cell breathing: initial results on the oxygen starvation problem. IEEE Control. Syst. Mag. **24**(2), 30–46 (2004)
44. Ramos-Paja, C.A., Giral, R., Martinez-Salamero, L., Romano, J., Romero, A., Spagnuolo, G.: A PEM fuel-cell model featuring oxygen-excess-ratio estimation and power-electronics interaction. IEEE Trans. Ind. Electron. **57**(6), 1914–1924 (2010)
45. Ramos Paja, C.A., Romero Nevado, A., Giral Castillón, R., Martinez-Salamero, L., Sanchez Saenz, C.I.: Switching and linear power stages evaluation for PEM fuel cell emulation. Int. J. Circuit Theory Appl. **39**(5), 475–499 (2011)
46. Restrepo, C., Ramos-Paja, C.A., Giral, R., Calvente, J., Romero, A.: Fuel cell emulator for oxygen excess ratio estimation on power electronics applications. Comput. Electr. Eng. **38**(4), 926–937 (2012)
47. Shtessel, Y.B., Baev, S., Edwards, C., Spurgeon, S.: HOSM observer for a class of non-minimum phase causal nonlinear mimo systems. IEEE Trans. Autom. Control. **55**(2), 543–548 (2010)
48. Slotine, J.J.E., Li, W., et al.: Applied nonlinear control, vol. 199. Prentice Hall New Jersey (1991)
49. Song, R.H., Kim, C.S., Shin, D.R.: Effects of flow rate and starvation of reactant gases on the performance of phosphoric acid fuel cells. J. Power Sources **86**(1–2), 289–293 (2000)
50. Talj, R.J., Hissel, D., Ortega, R., Becherif, M., Hilairet, M.: Experimental validation of a PEM fuel-cell reduced-order model and a moto-compressor higher order sliding-mode control. IEEE Trans. Ind. Electron. **57**(6), 1906–1913 (2010)
51. Thawornkuno, C., Panjapornpon, C.: Estimation of water content in PEM fuel cell. Chiang Mai J. Sci. **35**(1), 212–220 (2008)
52. Utkin, V.I.: Sliding Modes in Control and Optimization. Springer, Berlin (1992)
53. Vepa, R.: Adaptive state estimation of a PEM fuel cell. IEEE Trans. Energy Convers. **27**(2), 457–467 (2012)
54. Xiao, B., Hu, Q., Zhang, Y.: Adaptive sliding mode fault tolerant attitude tracking control for flexible spacecraft under actuator saturation. IEEE Trans. Control. Syst. Technol. **20**(6), 1605–1612 (2012)
55. Yan, X.G., Spurgeon, S.K., Edwards, C.: State and parameter estimation for nonlinear delay systems using sliding mode techniques. IEEE Trans. Autom. Control **58**(4), 1023–1029 (2013)

Chapter 7
Sliding-Mode-Observer-Based Fault Diagnosis of PEMFC Systems

This chapter presents a fault diagnosis method for PEMFC systems taking into account a fault scenario of sudden air leak in the air supply manifold. Based on the control-oriented model proposed in the literature, an adaptive-gain SOSM observer is developed for observing the system states, where the adaptive law estimates the uncertain parameters.

The residual signal is computed online from comparisons between the oxygen excess ratio obtained from the system model and the observer system, respectively. Equivalent output error injection using the residual signal is able to reconstruct the fault signal, which is critical in both fuel cell control design and fault detection. The performance of the proposed observer is validated through an HIL simulator which consists of a commercial twin screw compressor and a real-time PEMFC emulation system. The experimental results illustrate the feasibility and effectiveness of the proposed approach for application to PEMFC systems.

7.1 Introduction

Modern complex industrial systems can not be operated safely without reliable fault diagnosis and isolation (FDI) schemes in place [55]. Such systems including PEM fuel cells are vulnerable to system failures or mechanical faults that can lead to catastrophic consequences. PEM fuel cells are electrochemical devices that convert the chemical energy of a reaction between hydrogen and oxygen into electricity. Among different kinds of fuel cells, PEM fuel cells are suitable for both stationary and automobile applications due to the ongoing development of PEM technology [4, 6]. However, these energy generation systems based on fuel cells are complex and several additional equipments are required to make the fuel cell work at the optimal operating point. For this reason, they are vulnerable to system failures or

© Springer Nature Switzerland AG 2020
J. Liu et al., *Sliding Mode Control Methodology in the Applications of Industrial Power Systems*, Studies in Systems, Decision and Control 249,
https://doi.org/10.1007/978-3-030-30655-7_7

mechanical faults that can cause the shutdown or permanent damage of the fuel cell. Thus, reliable FDI schemes for such systems are of great importance.

FDI is usually achieved by generating residual signals, obtained from the difference between the actual system outputs and their estimated values calculated from dynamic models. Such approach usually involves two steps: the first step is to decouple the faults of interest from uncertainties and the second step is to generate residual signals and detect faults by decision logic. Several practical techniques for these steps have been proposed in contemporary literature, for example geometric approaches [31], H_∞-optimization technique [16, 34], observer based approaches (e.g. adaptive observers [49, 52, 53], high gain observers [3, 48], unknown input observers [5, 15]). However, in active Fault Tolerant Control (FTC) systems [55], not only the fault needs to be detected and isolated but also needs to be estimated such that its effect can be compensated by reconfiguring the controller [1, 8]. Hence, there is need of fault reconstruction schemes which estimate the fault's shape and magnitude. The challenge here is that the FDI approach should be able to differentiate between the effects of faults of interest and from uncertainties.

For linear systems, this challenge has been addressed by so called robust FDI schemes which are insensitive to model uncertainties [11]. In this respect, significant developments have been made and existing conditions have been given [28]. Qiu and Gertler [34] proposed residual generator design schemes based on the H_∞-optimization technique. Then, the so-called H_-/H_∞ approach was considered for the design of residual generator for uncertain systems in [16], where the effect of the uncertainty on the residual was taken to be upper-bounded and the effect of the fault to be lower bounded. Hou and Muller [15] used UIO to decouple the uncertainty from the FDI scheme. Qiu and Gertler [34] proposed residual generator design schemes based on the H_∞-optimization technique. Then, the so-called H_-/H_∞ approach was considered for the design of residual generator for uncertain systems in [16], where the effect of the uncertainty on the residual was taken to be upper-bounded and the effect of the fault to be lower bounded. Compared with linear systems, nonlinear systems are much more complicated, which makes their study more difficult. Even though a generic solution to the FDI problem is still to be developed, a number of effective observer based FDI approaches have been reported. These include techniques based on disturbances decoupling [14], geometric approaches [31] and so on.

Sliding mode technique is known for its insensitivity to external disturbances, high accuracy and finite time convergence. SMOs have been widely used for fault reconstruction in the past two decades. Edwards et al. [7] proposed a fault reconstruction approach based on equivalent output error injection. In this method, the resulting residual signal can approximate the actuator fault to any required accuracy. Based on the work of Edwards et al. [7], Tan et al. [44] proposed a sensor fault reconstruction method for well-modeled linear systems through the LMI technique. This approach is of less practical interest, as there is no explicit consideration of disturbance or uncertainty. To overcome this, the same authors [45] proposed an FDI scheme for a class of linear systems with uncertainty, using LMI for minimizing the L_2 gain between the uncertainty and the fault reconstruction signal. It should be noted that only linear systems are considered in the above works and only few works have been reported for nonlinear systems. Jiang et al. [18] proposed an SMO-based fault estimation approach for a class of nonlinear systems with uncertainties. Yan et al. [50] proposed a precise fault reconstruction scheme, based on equivalent output error injection, for a class of nonlinear systems with uncertainty. A sufficient condition based on LMI is presented for the existence and stability of a robust SMO. The limitation is that requires a strong structural condition of the distribution associated with uncertainties. Later, this structural constraint was relaxed by the same authors [51], where the fault distribution vector and the structure matrix of the uncertainty are allowed to be functions of the system's output and input. However, these works require that the bounds of the uncertainties and/or faults are *known*. Although in the work of Yan and Edwards [51], the requirement on the bound of uncertainty is removed, but it still needs to know the bound of the fault signal.

One of the main problems in the PEM fuel cell operation is the so-called oxygen starvation phenomenon during fast load variation. Accurate regulation of the oxygen excess ratio is required in order to avoid oxygen starvation [32]. This is a challenging task which comes from two aspects, on one hand, it is difficult to measure the oxygen excess ratio value, on the other hand, the fuel cell systems suffer from various faults, such as sudden air leak in the air supply manifold. Hence, from the point view of FTC, only FDI is not enough, the fault signal should be reconstructed and then its effect on the system performance can be compensated during active FTC design.

During the last decades, different kinds of model based techniques have been widely studied in the areas of FDI, health monitoring and complex industrial systems [38, 47]. Sliding mode based approach is one of the most attractive techniques due to its robustness against external disturbances, high accuracy and fast convergence [12, 23]. Several SMO-based FDI approaches have been proposed for linear systems [7, 45], but only few works have been reported for nonlinear systems, especially for nonlinear uncertain systems [12, 51]. However, first-order sliding mode algorithms are employed in the above works that require low pass filters to generate the output injection signals and induce undesirable chattering effects. Furthermore, the employment of low pass filters will introduce some delays which results in inaccurate estimates or even instability of the system. In recent years, HOSM technique has been widely studied due to the reasons that it does not require any low pass filters while keeping all the good properties of the standard sliding mode [25, 27]. This technique can also be used to alleviate the chattering effect because of its continuous output signal.

From the application point of view, several observer based FDI approaches have been studies for the fuel cell systems. Arcak et al. [2] developed an adaptive observer for hydrogen partial pressure estimation based on the fuel cell voltage. Görgün et al. [13] developed an algorithm for estimating partial pressures and the membrane water content in PEMFCs based on the resistive cell voltage drop. This algorithm has incorporated two adaptive observers for hydrogen and oxygen partial pressures, adapted from the work of Arcak et al. [2]. However, both of the above works lack robustness against the fuel cell voltage's measurement noise and the internal model relies upon unmeasurable values. Ingimundarson et al. [17] proposed a model based estimation approach to hydrogen leak detection in PEMFCs without the use of relative humidity sensors. Escobet et al. [9] designed a fault diagnosis methodology for PEMFCs where the residuals are generated from the differences between the PEMFC simulator included with a set of typical faults and a non-faulty fuel cell model. More recently, LPV observer was proposed by Lira et al. [26] for the application to fault detection in PEMFCs, where the stack current was taken as the scheduling variable. It should be noted that most of the cited papers are based on model linearization around pre-defined operating points of the system, depending upon the operating conditions such as temperature, humidity and air flow. Moreover, the calculation of Jacobian matrix of complex models like fuel cells are time consuming, and therefore difficult in real-time implementation. Several efficient control strategies have been designed in order to track the optimum operating points [19, 21, 56]. In the aim to assure optimal operating conditions, state estimation, parameter identification and fault reconstruction are going to play an important role. The state observers serve as a replacement of physical sensors for obtaining unmeasurable quantities such as oxygen and nitrogen partial pressures in the cathode side. Parameter identification is important for online monitoring procedures. The need for complete reconstruction of the faults not only facilitates fault diagnosis but also plays an important role in the enhancement of system robustness properties. Furthermore, state, parameter and fault estimations might be used for control purposes to maintain the fuel cell in an optimal operating points [10]. As is known the major obstacle to the implementability of these controllers, is the absence of reliable measurements of oxygen and nitrogen partial pressures in the cathode side. Therefore, the problems of state estimation and fault reconstruction arise due to incomplete knowledge of the parameters and states of the system. The auxiliary elements (e.g., valves, compressor, sensors) make the fuel cells vulnerable to several faults [9]. Moreover, the models used in [26, 32] are obtained through a Jacobian linearization of the PEMFC nonlinear dynamic model around its pre-established optimal operating point. As is known that the fuel cell system exhibits highly nonlinear dynamics and a fuel cell's efficiency is highly dependent on the various parameters such as operating conditions, such as temperature, humidity and air flow. A nonlinear four state dynamic model of the air-feed system proposed by Suh and Stefanopoulou [42] is chosen for state estimation and fault reconstruction. This model has been validated in the complete operating range of a fuel cell. A set of auxiliary elements (valves, compressor, sensors, etc.) are needed to make the fuel cell work at the optimal operating point. Thus, the problems of state estimation and fault reconstruction arise because of incomplete knowledge of the parameter and states of the system.

In this Chapter, state estimation, parameter identification and fault reconstruction are studied simultaneously for a class of nonlinear uncertain systems with Lipschitz nonlinearities. This class of systems sufficiently defines the nonlinear dynamics of the PEMFCs. The approach involves a simple adaptive update law and an adaptive-gain SOSM observer. The uncertain parameters are estimated and then injected into an adaptive-gain SOSM observer, which maintains a sliding motion even in the presence of fault signals. Finally, once the sliding motion is achieved, the equivalent output error injection can be obtained directly. Thus, the fault signals are reconstructed based on this information [7, 41]. The proposed SOSM algorithm combines the nonlinear term of the STA and a linear term [30]. The behavior of the STA near the origin is significantly improved compared with the linear case. Conversely, the additional linear term improves the behavior of the STA when the states are far from the origin. Therefore, the proposed algorithm inherits the best properties of both the linear and the nonlinear terms.

The proposed fault reconstruction scheme is evaluated by implementing on an instrumented HIL test bench, which consists of a real commercial twin screw compressor and a real-time fuel cell emulation system [29]. This considered model has been experimentally validated on a 33-kW PEM fuel cell in a wide operating range with less than 5% relative error [43]. A fault scenario, i.e., sudden air leak in the air supply manifold is considered. Its effect is simulated with an increment of supply manifold outlet flow constant, which presents a mechanical failure in the air circuit resulting in an abnormal air flow [9]. The system states (nitrogen partial pressure and compressor speed) are estimated successfully. The stack current, considered as an uncertain parameter, is estimated through an adaptive update law. This eliminates the need of an extra current sensor. The fault scenario considered in this study is a suddenly air leak in the air supply manifold, which results a change in the outlet air flow in the supply manifold. It is reconstructed faithfully through analyzing the information, which is obtained on-line from comparisons between the measurements from the sensor installed in the real system and the outputs of the observer system. The phenomenon so-called oxygen starvation is monitored through the estimation of a performance variable oxygen excess ratio. Since it will cause hot-spots on the membrane, resulting in permanent damage to the fuel cell if this ratio is less than 1 for a long time. The robustness of this observer against measurement noise and parameter variations is also validated experimentally.

7.2 Dynamic Modeling of PEMFC Systems

A typical PEM fuel cell system is shown in Fig. 5.1 which consists of four main subsystems, i.e., the air feed subsystem, the hydrogen supply subsystem, the humidify subsystem and the cooling subsystem. In order to achieve high efficient operation, a set of auxiliary elements (valves, compressor, sensors, etc.) are needed to make the fuel cell work at the optimal operating point. In this study, we will focus on the controller design of the air compressor for the air feed subsystem. The air compressor used to supply the oxygen to the cathode side is the core component and can consumes the power generated by the fuel cell up to 30% [46]. Therefore, efficient control of the air compressor is critical for the whole system and effects the system's efficiency directly. A typical PEM fuel cell polarization curve is shown in Fig. 7.1.

As is widely known, the dynamics of the PEM fuel cell system are with highly non-linearities. Therefore, suitable control-orient model taking into account the dynamic behaviors of the cathode partial pressure dynamics, the air supply manifold dynamics and the compressor dynamics is needed for the controller design. Some assumptions

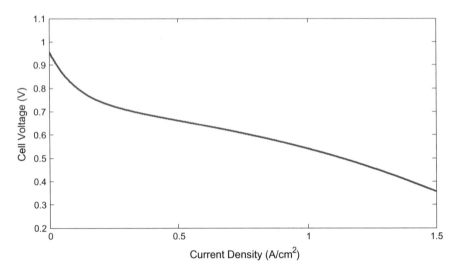

Fig. 7.1 Typical fuel cell voltage

are made to simplify the nonlinear model of the fuel cell system while keeping the dynamic behaviors of the air-feed subsystem [33]. Mainly, it is assumed that the temperature of the cathode inlet flow is regulated to a constant value through a heat exchanger. This is reasonable because the response time of the stack temperature is slow [22]. The relative humidity of both anode and cathode sides of the fuel cells are regulated to the desired relative humidity through an instantaneous humidifier. The hydrogen pressure in the anode side is regulated to follow the cathode pressure by the anode valve. Only vapor phase is considered inside the cathode and extra water in liquid phase is removed from the channels. The compressor motor current dynamics are neglected because the electrical time constant is very small as compared to the mechanical dynamics [36].

7.2.1   State-Space Representation of PEMFC Modeling

In view of (4.29), and define the state variables $x = [x_1, x_2, x_3, x_4]^T = [p_{O_2}, p_{N_2}, \omega_{cp}, p_{sm}]^T$ and $X = x_1 + x_2 + b_2$. Then, the nonlinear dynamics of the PEMFC system are expressed by the following equations:

$$\dot{x} = F(x) + G \cdot u + \Psi \cdot \xi, \tag{7.1}$$

where

$$F(x) = \begin{bmatrix} b_1(x_4 - X) - x_1 f_1(x_1, x_2) \\ b_1(x_4 - X) - x_2 f_1(x_1, x_2) \\ -c_9 x_3 - \frac{b_{10}}{x_3}\left[\left(\frac{x_4}{b_{11}}\right)^{b_{12}} - 1\right] f_2(x_3, x_4) \\ f_3(x_4)\left[f_2(x_3, x_4) - b_{16}(x_4 - X)\right] \end{bmatrix},$$

$$G = \begin{bmatrix} 0 & 0 & b_{13} & 0 \end{bmatrix}^T, \quad \Psi = \begin{bmatrix} -b_7 & 0 & 0 & 0 \end{bmatrix}^T,$$

(7.2)

where $f_1(x_1, x_2) := \frac{b_3}{b_4 x_1 + b_5 x_2 + b_6} W_{ca,out}$, $f_2(x_3, x_4) := W_{cp}$ and $f_3(x_4) := b_{14}$ $\left(1 + b_{15}\left[\left(\frac{x_4}{b_{11}}\right)^{b_{12}} - 1\right]\right)$.

The stack current $\xi := I_{st}$ is considered as the external disturbance and the control input $u := v_{cm}$ is the motor's input voltage. The outputs and performance variables of the system are given by:

$$y = \begin{bmatrix} y_1 & y_2 & y_3 \end{bmatrix}^T = \begin{bmatrix} p_{sm} & W_{cp} & V_{st} \end{bmatrix}^T, \quad z = \begin{bmatrix} z_1 & z_2 \end{bmatrix}^T = \begin{bmatrix} P_{net} & \lambda_{O_2} \end{bmatrix}^T,$$

(7.3)

where P_{net} and λ_{O_2} are fuel cell net power and oxygen excess ratio, respectively. The fuel cell net power P_{net} is the difference between the power produced by the stack P_{st} and the power consumed by the compressor. Thus, the net power can be expressed as:

$$P_{net} = P_{st} - P_{cp},$$

(7.4)

where $P_{st} = I_{st} V_{st}$ and $P_{cp} = \tau_{cm} \omega_{cp}$ are the stack power and compressor power, respectively. The oxygen excess ratio λ_{O_2} is defined by the following equation:

$$\lambda_{O_2} = \frac{W_{O_2,in}}{W_{O_2,react}} = \frac{b_{18}(p_{sm} - p_{ca})}{b_{19} I_{st}}.$$

(7.5)

The parameters b_i, $i \in \{1, \dots, 19\}$ are defined in Table 4.3. More details on this model are available in [32, 42].

Remark 7.1 As is well known that a set of auxiliary are needed to make the fuel cell work at its optimal operating conditions. Particularly, due to it is essential to regulate the oxygen excess ratio close to 2 in the presence of fast load variation and fault scenario [37, 39]. The level of λ_{O_2} is critical because fast load demand will result sudden decrease of the oxygen flow rate. On one hand, once the value of oxygen excess ratio decreases to a critical value (normally less than 1), oxygen starvation phenomenon which leads to the FC degradation occurs. On the other hand, higher value improves the fuel cell stack power but also result in higher power consumption of the air compressor. Therefore, the problems of oxygen excess ratio estimation and fault reconstruction arise due to the reasons of safety and high efficiency.

7.3   Nonlinear Observer Design for PEMFC System

In this Section, we consider the problem of an adaptive SOSM observer for a class of nonlinear systems, in which the uncertain parameter is estimated with the help of an adaptive law. Then, an SOSM observer with gain adaptation is developed using the estimated parameter. Finally, based on the adaptive SOSM observer, a fault reconstruction method which can be implemented online is proposed.

7.3.1   Problem Formulation

Consider the following nonlinear system

$$\dot{x} = Ax + g(x, u) + \phi(y, u)\theta + \omega(y, u)f(t),$$
$$y = Cx, \tag{7.6}$$

where the matrix $A = \begin{bmatrix} A_1 & A_2 \\ A_3 & A_4 \end{bmatrix}$, $x \in \mathbb{R}^n$ is the system state vector, $u(t) \in \mathcal{U} \subset \mathbb{R}^m$ is the control input, $y \in \mathcal{Y} \subset \mathbb{R}^p$ is the output vector, $g(x, u) \in \mathbb{R}^n$ is Lipschitz continuous, $\phi(y, u) \in \mathbb{R}^{n \times q}$ and $\omega(y, u) \in \mathbb{R}^{n \times r}$ are assumed to be some smooth and bounded functions with $p \geq q + r$. The unknown parameter vector $\theta \in \mathbb{R}^q$ is assumed to be constant and $f(t) \in \mathbb{R}^r$ is a smooth fault signal vector, which satisfies

$$\|f(t)\| \leq \rho_1, \quad \|\dot{f}(t)\| \leq \rho_2, \tag{7.7}$$

where ρ_1, ρ_2 are some positive constants that might be known or unknown.

Assume that (A, C) is an observable pair, and there exists a linear coordinate transformation $z = Tx = \begin{bmatrix} I_p & 0 \\ -H_{(n-p) \times p} & I_{n-p} \end{bmatrix} x = \begin{bmatrix} z_1^T & z_2^T \end{bmatrix}^T$, with $z_1 \in \mathbb{R}^p$ and $z_2 \in \mathbb{R}^{n-p}$, such that

- $TAT^{-1} = \begin{bmatrix} A_{11} & A_{12} \\ A_{21} & A_{22} \end{bmatrix}$, where the matrix $A_{22} = A_4 - HA_2 \in \mathbb{R}^{(n-p) \times (n-p)}$ is Hurwitz stable.
- $CT^{-1} = \begin{bmatrix} I_p & 0 \end{bmatrix}$, where $I_p \in \mathbb{R}^{p \times p}$ is an identity matrix.

Assumption 7.1 There exists a function $\omega_1(y, u)$ such that

$$T\omega(y, u) = \begin{bmatrix} \omega_1(y, u) \\ 0 \end{bmatrix}, \tag{7.8}$$

where $\omega_1(y, u) \in \mathbb{R}^{p \times r}$.

System (7.6) is described by the following equations in the new coordinate,

$$\dot{z} = TAT^{-1}z + Tg(T^{-1}z, u) + T\phi(y, u)\theta + T\omega(y, u)f(t),$$
$$y = CT^{-1}z. \tag{7.9}$$

By reordering the state variables, the system (7.9) can be rewritten as

$$\dot{y} = A_{11}y + A_{21}z_2 + g_1(z_2, y, u) + \phi_1(y, u)\theta + \omega_1(y, u)f(t),$$
$$\dot{z}_2 = A_{22}z_2 + A_{21}y + g_2(z_2, y, u) + \phi_2(y, u)\theta,$$
$$y = z_1, \tag{7.10}$$

where

$$T\phi(y, u) = \begin{bmatrix} \phi_1(y, u) \\ \phi_2(y, u) \end{bmatrix}, \quad Tg(T^{-1}z, u) = \begin{bmatrix} g_1(z_2, y, u) \\ g_2(z_2, y, u) \end{bmatrix}, \tag{7.11}$$

$\phi_1(\cdot, \cdot) : \mathbb{R}^p \times \mathbb{R}^m \rightarrow \mathbb{R}^p$, $\phi_2(\cdot, \cdot) : \mathbb{R}^p \times \mathbb{R}^m \rightarrow \mathbb{R}^{n-p}$, $g_1(\cdot, \cdot, \cdot) : \mathbb{R}^p \times \mathbb{R}^{n-p} \times \mathbb{R}^m \rightarrow \mathbb{R}^p$, $g_2(\cdot, \cdot, \cdot) : \mathbb{R}^p \times \mathbb{R}^{n-p} \times \mathbb{R}^m \rightarrow \mathbb{R}^{n-p}$.

Remark 7.2 The Assumption 7.1 is a structural constraint on the fault distribution $\omega(y, u)$. It should be noted that there are no such structural constraints on the uncertain parameter distribution $\phi(\cdot, \cdot)$. The structure of (7.10) is more general than that proposed in [51], since there exists Lipschitz nonlinearities $g_1(z_2, y, u), g_2(z_2, y, u)$ which depend on the states.

7.3.2 Adaptive SOSM Observer Design

Now, we will develop an adaptive SOSM observer for the system (7.10), which consists of an adaptive update law and an adaptive-gain SOSM observer, to estimate the uncertain parameter and system state variables, respectively. The basic assumption on the system (7.10) is as follows:

Assumption 7.2 ([51]) There exists a nonsingular matrix $\bar{T} \in \mathbb{R}^{p \times p}$, such that

$$\bar{T} \begin{bmatrix} \phi_1(y, u) & \omega_1(y, u) \end{bmatrix} = \begin{bmatrix} \Phi_1(y, u) & 0_{q \times r} \\ 0_{r \times q} & \Phi_2(y, u) \end{bmatrix}, \tag{7.12}$$

where $\Phi_1(y, u) \in \mathbb{R}^{q \times q}$, $\Phi_2(y, u) \in \mathbb{R}^{r \times r}$ are both nonsingular matrices and bounded in $(y, u) \in \mathcal{Y} \times \mathcal{U}$.

Remark 7.3 The main limitation in Assumption 7.2 is that the matrices $[\phi_1(y, u), \omega_1(y, u)]$ must be block-diagonalizable by elementary row transformations [51]. For the sake of simplicity, the case of only one fault signal and one uncertain parameter is considered ($q = r = 1$).

Let $z_y = \bar{T}y$, where $\bar{T}$ is defined in Assumption 7.2. Then, the system (7.10) can be described by

$$
\begin{aligned}
\dot{z}_y &= \bar{T}A_{11}y + \bar{T}A_{21}z_2 + \bar{T}g_1(y, z_2, u) + \begin{bmatrix} \Phi_1(y, u) \\ 0 \end{bmatrix}\theta + \begin{bmatrix} 0 \\ \Phi_2(y, u) \end{bmatrix}f(t), \\
\dot{z}_2 &= A_{22}z_2 + A_{21}y + g_2(z_2, y, u) + \phi_2(y, u)\theta, \\
y &= \bar{T}^{-1}z_y,
\end{aligned}
\tag{7.13}
$$

where

$$
\bar{T} \cdot A_{11} = \begin{bmatrix} \bar{A}_{11} \\ \bar{A}_{12} \end{bmatrix}, \quad \bar{T} \cdot A_{12} = \begin{bmatrix} \bar{A}_{21} \\ \bar{A}_{22} \end{bmatrix}, \quad \bar{T} \cdot g_1(y, z_2, u) = \begin{bmatrix} W_{g_1}(y, z_2, u) \\ W_{g_2}(y, z_2, u) \end{bmatrix}.
\tag{7.14}
$$

Let define $z_y = \begin{bmatrix} z_{y_1}, z_{y_2} \end{bmatrix}^\mathrm{T}$, with $z_{y_1} \in \mathbb{R}^q$, $z_{y_2} \in \mathbb{R}^r$. Then, in view of (7.13) and (7.14), we can obtain

$$
\begin{aligned}
\dot{z}_{y_1} &= \bar{A}_{11}y + \bar{A}_{21}z_2 + W_{g_1}(y, z_2, u) + \Phi_1(y, u)\theta, \\
\dot{z}_{y_2} &= \bar{A}_{12}y + \bar{A}_{22}z_2 + W_{g_2}(y, z_2, u) + \Phi_2(y, u)f(t), \\
\dot{z}_2 &= A_{21}y + A_{22}z_2 + g_2(y, z_2, u) + \phi_2(y, u)\theta, \\
y &= T^{-1}\begin{bmatrix} z_{y_1} & z_{y_2} \end{bmatrix}^\mathrm{T}.
\end{aligned}
\tag{7.15}
$$

The adaptive SOSM observer is represented by the following dynamical system

$$
\begin{aligned}
\dot{\hat{z}}_{y_1} &= \bar{A}_{11}y + \bar{A}_{21}\hat{z}_2 + W_{g_1}(y, \hat{z}_2, u) + \Phi_1(y, u)\hat{\theta} + \mu(e_{y_1}), \\
\dot{\hat{z}}_{y_2} &= \bar{A}_{12}y + \bar{A}_{22}\hat{z}_2 + W_{g_2}(y, \hat{z}_2, u) + \mu(e_{y_2}), \\
\dot{\hat{z}}_2 &= A_{21}y + A_{22}\hat{z}_2 + g_2(y, \hat{z}_2, u) + \phi_2(y, u)\hat{\theta},
\end{aligned}
\tag{7.16}
$$

where $\mu(\cdot)$ is calculated by the SOSM algorithm

$$
\mu(s) = \lambda(t)|s|^{\frac{1}{2}}\mathrm{sign}(s) + \alpha(t)\int_0^t \mathrm{sign}(s)d\tau + k_\lambda(t)s + k_\alpha(t)\int_0^t sd\tau,
\tag{7.17}
$$

with the adaptive gains $\lambda(t)$, $\alpha(t)$, $k_\lambda(t)$ and $k_\alpha(t)$ which will be determined later on.

The observation errors are defined as $e_{y_1} = z_{y_1} - \hat{z}_{y_1}$, $e_{y_2} = z_{y_2} - \hat{z}_{y_2}$, $e_2 = z_2 - \hat{z}_2$, and $\tilde{\theta} = \theta - \hat{\theta}$. The estimate of θ, denoted by $\hat{\theta}$, is given by the following adaptive law

$$
\dot{\hat{\theta}} = -K(y, u)\left(\bar{A}_{11}y + \bar{A}_{21}\hat{z}_2 + W_{g_1}(y, \hat{z}_2, u) + \Phi_1(y, u)\hat{\theta} - \dot{z}_{y_1} \right),
\tag{7.18}
$$

where $K(y, u)$ is a positive design matrix which will be defined later.

Remark 7.4 It can be seen that the adaptive law (7.18) depends upon $\dot{z}_{y_1}$. The adaptive-gain SOSM algorithm (7.17) can be used to estimate the time derivative of z_{y_1} in finite time, without requiring the bound of $|\ddot{z}_{y_1}|$. The differentiator has the following form

$$
\begin{aligned}
\dot{z}_0 &= -\lambda_0 \sqrt{L(t)} |z_0 - z_{y_1}|^{\frac{1}{2}} \operatorname{sign}(z_0 - z_{y_1}) - k_{\lambda_0} L(t)(z_0 - z_{y_1}) + z_1, \\
\dot{z}_1 &= -\alpha_0 L(t) \operatorname{sign}(z_0 - z_{y_1}) - k_{\alpha_0} L^2(t)(z_0 - z_{y_1}),
\end{aligned} \tag{7.19}
$$

where z_0 and z_1 are the real-time estimations of z_{y_1} and $\dot{z}_{y_1}$ respectively. The parameters of the differentiator λ_0, k_{λ_0}, α_0, k_{α_0} and the time varying function $L(t)$ are designed as (7.33) and (7.32), respectively.

Subtracting (7.16) from (7.15), the error dynamical equation is described by

$$
\begin{aligned}
\dot{e}_2 &= A_{22}e_2 + \tilde{g}_2(y, z_2, \hat{z}_2, u) + \phi_2(y, u)\tilde{\theta}, & (7.20) \\
\dot{\tilde{\theta}} &= -K(y, u)\left(\bar{A}_{21}e_2 + \tilde{W}_{g_1}(y, z_2, \hat{z}_2, u) + \Phi_1(y, u)\tilde{\theta}\right), & (7.21) \\
\dot{e}_{y_1} &= -\mu(e_{y_1}) + \bar{A}_{21}e_2 + \Phi_1(y, u)\tilde{\theta} + \tilde{W}_{g_1}(y, z_2, \hat{z}_2, u), & (7.22) \\
\dot{e}_{y_2} &= -\mu(e_{y_2}) + \bar{A}_{22}e_2 + \Phi_2(y, u)f(t) + \tilde{W}_{g_2}(y, z_2, \hat{z}_2, u), & (7.23)
\end{aligned}
$$

where

$$
\begin{aligned}
\tilde{g}_2(y, z_2, \hat{z}_2, u) &= g_2(y, z_2, u) - g_2(y, \hat{z}_2, u) \\
\tilde{W}_{g_1}(y, z_2, \hat{z}_2, u) &= W_{g_1}(y, z_2, u) - W_{g_1}(y, \hat{z}_2, u) \\
\tilde{W}_{g_2}(y, z_2, \hat{z}_2, u) &= W_{g_2}(y, z_2, u) - W_{g_2}(y, \hat{z}_2, u).
\end{aligned}
$$

Some assumptions are imposed upon the error systems (7.20)–(7.23).

Assumption 7.3 The known nonlinear terms $g_2(y, z_2, u)$, $W_{g_1}(y, z_2, u)$ and $W_{g_2}(y, z_2, u)$ are Lipschitz continuous with respect to z_2

$$
\begin{aligned}
\|g_2(y, z_2, u) - g_2(y, \hat{z}_2, u)\| &\leq \gamma_2 \|z_2 - \hat{z}_2\|, \\
\|W_{g_1}(y, z_2, u) - W_{g_1}(y, \hat{z}_2, u)\| &\leq \gamma_{g_1} \|z_2 - \hat{z}_2\|, & (7.24) \\
\|W_{g_2}(y, z_2, u) - W_{g_2}(y, \hat{z}_2, u)\| &\leq \gamma_{g_2} \|z_2 - \hat{z}_2\|,
\end{aligned}
$$

where γ_{g_1}, γ_{g_2} and γ_2 are some known Lipschitz constants for $W_{g_1}(y, z_2, u)$, $W_{g_2}(y, z_2, u)$ and $g_2(y, z_2, u)$ respectively [54].

Assumption 7.4 Assume that the Huritwz matrix A_{22} satisfies the following Riccati equation

$$
A_{22}^T P_1 + P_1 A_{22} + \gamma_2^2 P_1 P_1 + (2 + \varepsilon)I_{n-p} = 0, \tag{7.25}
$$

which has a symmetric positive-definite solution P_1 for some $\varepsilon > 0$ [35].

Assumption 7.5 Assume that the positive design matrix $K(y, u)$ satisfies the following equation

$$K(y, u)\Phi_1(y, u) + \Phi_1^{\mathrm{T}}(y, u)K^{\mathrm{T}}(y, u) - \gamma_{g_1}^2 K(y, u)K^{\mathrm{T}}(y, u) = \epsilon I_q, \quad (7.26)$$

for some $\epsilon > 0$.

Remark 7.5 The Assumption 7.3 means that the nonlinear terms $g_2(y, z_2, u)$, $W_{g_1}(y, z_2, u)$ and $W_{g_2}(y, z_2, u)$ are Lipschitz with respect to the variable z_2. The Assumption 7.4 means that the Hurwitz matrix A_{22} satisfies the Riccati equation (7.25) for some small $\varepsilon > 0$. The Assumption 7.5 means that the design matrix $K(y, u)$ must be designed such that $K(y, u)\Phi_1(y, u) + \Phi_1^{\mathrm{T}}(y, u)K^{\mathrm{T}}(y, u)$ is positive definite in $(y, u) \in \mathcal{Y} \times \mathcal{U}$.

Now, we will first consider the stability of the error systems (7.20) and (7.21).

Theorem 7.1 *Consider the systems (7.20) and (7.21) satisfying the Assumptions 7.3–7.5. Then, the error systems (7.20) and (7.21) are exponentially stable, if for any $(y, u) \in \mathcal{Y} \times \mathcal{U}$, the following matrix*

$$Q_1 = \begin{bmatrix} \varepsilon I_{n-p}, & P_1\phi_2(y, u) - \bar{A}_{21}^{\mathrm{T}} K^{\mathrm{T}}(y, u) \\ \phi_2^{\mathrm{T}}(y, u)P_1 - K(y, u)\bar{A}_{21}, & \epsilon I_q \end{bmatrix} \quad (7.27)$$

is positive definite.

Proof A candidate Lyapunov function is chosen as

$$V_1(e_2, \tilde{\theta}) = e_2^{\mathrm{T}} P_1 e_2 + \tilde{\theta}^{\mathrm{T}}\tilde{\theta} \quad (7.28)$$

and the time derivative of V_1 along the solution of the system (7.20) and (7.21) is given by

$$\begin{aligned}
\dot{V}_1 &\le e_2^{\mathrm{T}}(A_{22}^{\mathrm{T}} P_1 + P_1 A_{22} + \gamma_2^2 P_1 P_1 + 2I_{n-p})e_2 \\
&\quad + 2e_2^{\mathrm{T}}\left(P_1\phi_2(y, u) - \bar{A}_{21}^{\mathrm{T}} K^{\mathrm{T}}(y, u)\right)\tilde{\theta} - \epsilon\tilde{\theta}^{\mathrm{T}}\tilde{\theta} \\
&= -\begin{bmatrix} e_2^{\mathrm{T}} & \tilde{\theta}^{\mathrm{T}} \end{bmatrix} Q_1 \begin{bmatrix} e_2 \\ \tilde{\theta} \end{bmatrix}.
\end{aligned} \quad (7.29)$$

Hence, the conclusion follows given that Q_1 is positive definite in $(y, u) \in \mathcal{Y} \times \mathcal{U}$.
∎

Assumption 7.6 It is assumed that $\|\Phi_1(y, u)\|$, $\|\Phi_2(y, u)\|$ are bounded in $(y, u) \in \mathcal{Y} \times \mathcal{U}$.

Remark 7.6 Theorem 7.1 shows that $\lim_{t\to\infty} e_2(t) = 0$ and $\lim_{t\to\infty} \tilde{\theta}(t) = 0$. Consequently, the errors e_2, $\tilde{\theta}$ and its derivatives $\dot{e}_2$, $\dot{\tilde{\theta}}$ are bounded. Under Assumptions 7.3 and 7.6, the time derivatives of the nonlinear terms in the error dynamics (7.22) and (7.23) are also bounded

$$\left\| \frac{d}{dt} \left(\bar{A}_{22} e_2 + \Phi_2(y, u) f(t) + \tilde{W}_{g_2}(y, z_2, \hat{z}_2, u) \right) \right\| \leq \chi_1,$$

$$\left\| \frac{d}{dt} \left(\bar{A}_{21} e_2 + \Phi_1(y, u) \tilde{\theta} + \tilde{W}_{g_1}(y, z_2, \hat{z}_2, u) \right) \right\| \leq \chi_2, \tag{7.30}$$

where χ_1 and χ_2 are some *unknown* positive constants.

In what follows, the objective is to prove the finite time convergence of the system (7.22) and (7.23).

Theorem 7.2 *Assume that (7.30) holds and the adaptive gains* $\lambda(t)$, $\alpha(t)$, $k_\lambda(t)$ *and* $k_\alpha(t)$ *in the SOSM algorithm (7.17) are formulated as*

$$\lambda(t) = \lambda_0 \sqrt{L(t)}, \quad \alpha(t) = \alpha_0 L(t),$$

$$k_\lambda(t) = k_{\lambda_0} L(t), \quad k_\alpha(t) = k_{\alpha_0} L^2(t), \tag{7.31}$$

where $\lambda_0, \alpha_0, k_{\lambda_0}$ *and* k_{α_0} *are positive constants and* $L(t)$ *is a positive, time-varying, scalar function. The adaptive law of the time-varying function* $L(t)$ *is given by*

$$\dot{L}(t) = \begin{cases} k, & \text{if } |e_{y_i}| \neq 0, \ i = \{1, 2\}, \\ 0, & \text{else}, \end{cases} \tag{7.32}$$

where k *is an arbitrary positive value. Then, the trajectories of the error system (7.22) and (7.23) converge to zero in finite time, if* $\lambda_0, \alpha_0, k_{\lambda_0}$ *and* k_{α_0} *in (7.31) satisfy*

$$4\alpha_0 k_{\alpha_0} > 8k_{\lambda_0}^2 \alpha_0 + 9\lambda_0^2 k_{\lambda_0}^2. \tag{7.33}$$

Proof We only consider the proof for the system (7.23) and the proof for the system (7.22) takes the same way. The system (7.23) can be rewritten as

$$\dot{e}_{y_2} = -\lambda(t) |e_{y_2}|^{\frac{1}{2}} \text{sign}(e_{y_2}) - k_\lambda(t) e_{y_2} + \varphi,$$

$$\dot{\varphi} = -\alpha(t) \text{sign}(e_{y_2}) - k_\alpha(t) e_{y_2} + \varrho(t). \tag{7.34}$$

Under the condition (7.30), it follows that $\|\varrho(t)\| \leq \chi_2$, where χ_2 is an *unknown* positive constant.

A new state vector is introduced to represent the system (7.34) in a more convenient form for Lyapunov analysis.

$$\zeta = \begin{bmatrix} \zeta_1 \\ \zeta_2 \\ \zeta_3 \end{bmatrix} = \begin{bmatrix} L^{\frac{1}{2}}(t) |e_{y_2}|^{\frac{1}{2}} \text{sign}(e_{y_2}) \\ L(t) e_{y_2} \\ \varphi \end{bmatrix}. \tag{7.35}$$

Thus, the system (7.34) can be rewritten as

$$\dot{\zeta} = \frac{L(t)}{|\zeta_1|} \begin{bmatrix} -\frac{\lambda_0}{2} & 0 & \frac{1}{2} \\ 0 & -\lambda_0 & 0 \\ -\alpha_0 & 0 & 0 \end{bmatrix} \zeta + L(t) \begin{bmatrix} -\frac{k_{\lambda_0}}{2} & 0 & 0 \\ 0 & -k_{\lambda_0} & 1 \\ 0 & -k_{\alpha_0} & 0 \end{bmatrix} \zeta + \begin{bmatrix} \frac{\dot{L}}{2L(t)}\zeta_1 \\ \frac{\dot{L}}{2L(t)}\zeta_2 \\ \varrho(t) \end{bmatrix}. \tag{7.36}$$

Then, the following Lyapunov function candidate is introduced for the system (7.36)

$$V(\zeta) = \zeta^T P \zeta, \quad P = \frac{1}{2} \begin{bmatrix} 4\alpha_0 + \lambda_0^2 & \lambda_0 k_{\lambda_0} & -\lambda_0 \\ \lambda_0 k_{\lambda_0} & k_{\lambda_0}^2 + 2k_{\alpha_0} & -k_{\lambda_0} \\ -\lambda_0 & -k_{\lambda_0} & 2 \end{bmatrix}, \tag{7.37}$$

where the matrix P is symmetric positive definite due to the fact that its leading principle minors are all positive under the condition (7.33).

Taking the derivative of (7.37) yields

$$\dot{V} = -L(t)\left(\frac{\zeta^T \Omega_1 \zeta}{|\zeta_1|} + \zeta^T \Omega_2 \zeta\right) + \varrho(t)q_1\zeta + \frac{q_2\dot{L}(t)}{L(t)}P\zeta, \tag{7.38}$$

where $q_1 = \begin{bmatrix} -\lambda_0 & -k_{\lambda_0} & 2 \end{bmatrix}$, $q_2 = \begin{bmatrix} \zeta_1 & \zeta_2 & 0 \end{bmatrix}$ and

$$\Omega_1 = \frac{\lambda_0}{2} \begin{bmatrix} \lambda_0^2 + 2\alpha_0 & 0 & -\lambda_0 \\ 0 & 2k_{\alpha_0} + 5k_{\lambda_0}^2 & -3k_{\lambda_0} \\ -\lambda_0 & -3k_{\lambda_0} & 1 \end{bmatrix},$$

$$\Omega_2 = k_{\lambda_0} \begin{bmatrix} \alpha_0 + 2\lambda_0^2 & 0 & 0 \\ 0 & k_{\alpha_0} + k_{\lambda_0}^2 & -k_{\lambda_0} \\ 0 & -k_{\lambda_0} & 1 \end{bmatrix}. \tag{7.39}$$

It is easy to verify that Ω_1 and Ω_2 are positive definite matrices under the condition (7.33). Since $\lambda_{\min}(P)\|\zeta\|^2 \leq V \leq \lambda_{\max}(P)\|\zeta\|^2$, Eq. (7.38) can be rewritten as

$$\dot{V} \leq -L(t)\frac{\lambda_{\min}(\Omega_1)}{\lambda_{\max}^{\frac{1}{2}}(P)}V^{\frac{1}{2}} - L(t)\frac{\lambda_{\min}(\Omega_2)}{\lambda_{\max}(P)}V + \frac{\chi_2\|q_1\|_2}{\lambda_{\min}^{\frac{1}{2}}(P)}V^{\frac{1}{2}} + \frac{\dot{L}(t)}{2L(t)}\zeta^T Q\zeta, \tag{7.40}$$

where

$$Q = \begin{bmatrix} 4\alpha_0 + \lambda_0^2 + \lambda_0 k_{\lambda_0} + \frac{\lambda_0}{2} & 0 & 0 \\ 0 & \frac{\lambda_0 + k_{\lambda_0}}{2} & 0 \\ 0 & 0 & 2k_{\alpha_0}k_{\lambda_0}^2 + \lambda_0 k_{\lambda_0} + \frac{k_{\lambda_0}}{2} \end{bmatrix}. \tag{7.41}$$

Using (7.41), (7.40) becomes

$$\dot{V} \leq -(L(t)\gamma_1 - \gamma_2)V^{\frac{1}{2}} - \left(L(t)\gamma_3 - \gamma_4\frac{\dot{L}(t)}{L(t)}\right)V, \tag{7.42}$$

where

$$\gamma_1 = \frac{\lambda_{\min}(\Omega_1)}{\lambda_{\max}^{\frac{1}{2}}(P)}, \quad \gamma_2 = \frac{\chi_2 \|q_1\|_2}{\lambda_{\min}^{\frac{1}{2}}(P)}, \quad \gamma_3 = \frac{\lambda_{\min}(\Omega_2)}{\lambda_{\max}(P)}, \quad \gamma_4 = \frac{\lambda_{\max}(Q)}{2\lambda_{\min}(P)}. \tag{7.43}$$

Because $\dot{L}(t) \geq 0$ such that the terms $L(t)\gamma_1 - \gamma_2$ and $L(t)\gamma_3 - \gamma_4\frac{\dot{L}(t)}{L(t)}$ are positive in finite time. It follows from (7.43) that

$$\dot{V} \leq -c_1 V^{\frac{1}{2}} - c_2 V, \tag{7.44}$$

where c_1 and c_2 are positive constants. By the comparison principle [20], it gives that ζ converge to zero in finite time, i.e. $e_{y_2} = 0$ and $\dot{e}_{y_2} = 0$. This completes the proof. ∎

Remark 7.7 In view of practical implementation, the condition $|e_{y_i}| = 0$ in (7.32) can not be satisfied due to measurement noise and numerical approximations. In order to make the adaptive algorithm (7.32) practically implementable, one has to modify the condition $|e_{y_i}| = 0$ by dead-zone technique [40] as

$$\dot{L}(t) = \begin{cases} k, & \text{if } |e_{y_i}| \geq \tau, \\ 0, & \text{else}, \end{cases} \tag{7.45}$$

where τ is a sufficiently small positive value.

Theorems 7.1 and 7.2 have shown that systems (7.16) and (7.18) are an asymptotic state observer and uncertain parameter observer for the system (7.15) respectively. In the next Section, we will develop the fault reconstruction approach based on those two observers.

7.4   Fault Reconstruction

The fault signal $f(t)$ will be reconstructed based on the proposed observer by using an equivalent output error injection which can be obtained once the sliding surface is reached and maintained on it thereafter.

It follows from Theorem 7.2 that e_{y_2} and $\dot{e}_{y_2}$ in (7.23) are driven to zero in finite time. Thus, the equivalent output error injection can be obtained directly

$$\mu(e_{y_2}) = \bar{A}_{22}e_2 + \Phi_2(y, u)f(t) + \tilde{W}_{g_2}(y, z_2, \hat{z}_2, u). \tag{7.46}$$

From Assumption 7.2, $\Phi_2(y, u)$ is a bounded nonsingular matrix in $(y, u) \in \mathcal{Y} \times \mathcal{U}$, then the estimate of $f(t)$ can be constructed as

$$\hat{f}(t) = \Phi_2^{-1}(y, u)\mu(e_{y_2}). \tag{7.47}$$

Theorem 7.3 *Suppose that conditions of Theorems 7.1 and 7.2 are satisfied, then* $\hat{f}(t)$ *defined in* (7.47) *is a precise reconstruction of the fault* $f(t)$ *since*

$$\lim_{t \to \infty} \| f(t) - \hat{f}(t) \| = 0. \tag{7.48}$$

Proof It follows from (7.46) and (7.47) that

$$
\begin{aligned}
\| f(t) - \hat{f}(t) \| &= \| \Phi_2^{-1}(y, u)(\bar{A}_{22} e_2 + \tilde{W}_{g_2}) \| \\
&\le \| \Phi_2^{-1}(y, u) \bar{A}_{22} \| \| e_2 \| + \gamma_{g_2} \| \Phi_2^{-1}(y, u) \| \| e_2 \|.
\end{aligned}
\tag{7.49}
$$

It follows $\lim_{t \to \infty} \| f(t) - \hat{f}(t) \| = 0$ given that $\lim_{t \to \infty} \| e_2 \| = 0$. Hence, Theorem 7.3 is proven.   ∎

7.5 FDI of PEMFC System

Figure 5.1 shows a block diagram of a typical PEM fuel cell system in automotive applications, which consists of four major subsystems: the air feed subsystem, the hydrogen supply subsystem, the humidify subsystem and the cooling subsystem. The PEMFCs are supplied at the cathode with compressed air via a supply manifold. The compressor draws the air from the atmosphere directly. At the cathode exit, the air enters an outlet manifold that is open to the atmosphere. Several assumptions are made for the system. Mainly, the anode pressure is assumed to be well controlled and equal to the cathode pressure [33]. The input reactant flows are humidified in a consistent and rapid way and the high pressure compressed hydrogen is available. Vapor partial pressure in the stack is considered equal to the saturation pressure.

In view of the system (7.1), the state space representation of this model can be written is as follows

$$
\begin{aligned}
\dot{x}_1 &= -(c_1 + c_8)(x_1 - x_4) - \frac{c_3(x_1 - c_2)}{\kappa x_1} W_{ca,out} - c_7 \xi, \\
\dot{x}_2 &= c_8(x_4 - x_1) - \frac{c_3 x_2}{\kappa x_1} W_{ca,out}, \\
\dot{x}_3 &= -c_9 x_3 - \frac{c_{10}}{x_3} \left[\left(\frac{x_4}{c_{11}} \right)^{c_{12}} - 1 \right] W_{cp} + c_{13} u, \\
\dot{x}_4 &= c_{14} \left\{ 1 + c_{15} \left[\left(\frac{x_4}{c_{11}} \right)^{c_{12}} - 1 \right] \right\} \times \left[W_{cp} - c_{16}(x_4 - x_1) \right],
\end{aligned}
\tag{7.50}
$$

where $x_1 := p_{ca}$ is the cathode pressure, $x_2 := p_{N_2}$ is the nitrogen partial pressure, $x_3 := \omega_{cp}$ is the compressor speed and $x_4 := p_{sm}$ is the supply manifold pressure. $W_{ca,out}$ is the cathode flow rate which is a function of the cathode pressure $W_{ca,out} = k_{ca,out} \sqrt{p_{ca} - p_{atm}}$ and W_{cp} is the compressor flow rate which is a function of the angular speed of the compressor $W_{cp} = h_3(x_3) = c_{17}w_{cp}$ [29]. The stack current ξ is considered as an uncertain parameter θ. The control input u represents the motor's quadratic current component. The cathode and supply manifold pressures are assumed to be available for measurement. Thus, the system outputs are $y = [x_1 \ x_4]^T$. The system performance variable, oxygen excess ratio λ_{O_2} is defined as the ratio between the oxygen entering the cathode $W_{O_2,in}$ and the oxygen reacting in the fuel cell stack $W_{O_2,react}$:

$$\lambda_{O_2} = \frac{W_{O_2,in}}{W_{O_2,react}} = \frac{c_{18}(x_4 - x_1)}{c_{19}\xi}. \tag{7.51}$$

Due to the reasons of safety and high efficiency, it is typical to operate the stacks with this value equals 2 during step changes of load current demand [29]. It should be noted that positive deviations of λ_{O_2} above 2 imply lower efficiency, since excess oxygen supplied into the cathode will cause power waste, and negative deviations increase the probability of the starvation phenomena.

We consider a fault scenario: a sudden air leak in the air supply manifold. This fault is simulated with an increment Δc_{16} in the supply manifold outlet flow constant $c_{16} := k_{sm,out}$, which is translated into a change in the outlet air flow in the supply manifold $W_{sm,out} = (c_{16} + \Delta c_{16})(x_4 - x_1)$ [9, 26]. We assume that this fault appears after time $t = 50$ s, i.e., $\Delta c_{16} = 0.2c_{16}$. Thus, the fault signal $f(t)$ appears in the output channel x_4 is defined as

$$f(t) = \begin{cases} \Delta c_{16} \times (x_4 - x_1) \text{ kg/s,} & \text{if } t \geq 50 \text{ s,} \\ 0, & \text{else.} \end{cases} \tag{7.52}$$

In order to design the proposed observer for the fuel cell system. Let us define $z_{y_1} := x_1$, $z_{y_2} := x_4$, $z_2 := [x_2 \ x_3]^T$ and $\theta := \xi$. Then, system (7.50) is described as the form of (7.15)

$$\dot{z}_{y_1} = -(c_1 + c_8)(z_{y_1} - z_{y_2}) - \frac{c_3(z_{y_1} - c_2)W_{ca,out}}{\kappa z_{y_1}} - c_4\theta,$$

$$\dot{z}_{y_2} = c_{14}\left\{1 + c_{15}\left[\left(\frac{z_{y_2}}{c_{11}}\right)^{c_{12}} - 1\right]\right\} \times \left[h_3(D_2 z_2) - (c_{16} + \Delta c_{16})(z_{y_2} - z_{y_1})\right],$$

$$\dot{z}_2 = \underbrace{\begin{bmatrix} -H & 0 \\ 0 & -c_9 \end{bmatrix}}_{A_{22}} z_2 + \underbrace{\begin{bmatrix} -c_8 & c_8 \\ 0 & 0 \end{bmatrix}}_{A_{21}} y + \underbrace{\begin{bmatrix} D_1 z_2\left(H - \frac{c_3 W_{ca,out}}{\kappa z_{y_1}}\right) \\ -\frac{c_{10}}{D_2 z_2}\left[\left(\frac{z_{y_2}}{c_{11}}\right)^{c_{12}} - 1\right]h_3(D_2 z_2) + c_{13}u \end{bmatrix}}_{g_2(y,z_2,u)},$$

$$\tag{7.53}$$

where

$$D_1 = [1, 0], \quad D_2 = [0, 1], \quad \Phi_1(y, u) := -c_4, \quad \Phi_2(y, u) := c_5,$$

$$W_{g_1}(y, z_2, u) := -\frac{c_3(z_{y_1} - c_2)\psi(z_{y_1})}{\kappa z_{y_1}}, \quad \phi_2(y, u) = 0,$$

$$W_{g_2}(y, z_2, u) := c_{14}c_{15}\left(\frac{z_{y_2}}{c_{11}}\right)^{c_{12}}\left(h_3(D_2 z_2) - c_{16}(z_{y_2} - z_{y_1})\right),$$

and the design parameter H is chosen to satisfy the Riccati equation (7.25). The fault signal is weighted by c_5, modeled as

$$c_5 = -c_{14}\left\{1 + c_{15}\left[\left(\frac{z_{y_2}}{c_{11}}\right)^{c_{12}} - 1\right]\right\}. \tag{7.54}$$

The adaptive SOSM observer for the system (7.53) is designed as the form (7.16) and (7.18)

$$\dot{\hat{z}}_{y_1} = -(c_1 + c_8)(z_{y_1} - z_{y_2}) - \frac{c_3(z_{y_1} - c_2)}{\kappa z_{y_1}}W_{ca,out} - c_4\hat{\theta} + \mu(e_{y_1}),$$

$$\dot{\hat{z}}_{y_2} = c_{14}\left\{1 + c_{15}\left[\left(\frac{z_{y_2}}{c_{11}}\right)^{c_{12}} - 1\right]\right\} \times \left[h_3(D_2\hat{z}_2) - c_{16}(z_{y_2} - z_{y_1})\right] + \mu(e_{y_2}),$$

$$\dot{\hat{z}}_2 = A_{22}\hat{z}_2 + A_{21}y + g_2(y, \hat{z}_2, u), \tag{7.55}$$

and $\dot{\hat{\theta}} = -K\left(c_4\hat{\theta} + (c_1 + c_8)(z_{y_1} - z_{y_2}) + \frac{c_3(z_{y_1} - c_2)W_{ca,out}}{\kappa z_{y_1}} + \dot{z}_{y_1}\right)$, where $e_{y_1} = z_{y_1} - \hat{z}_{y_1}$, $e_{y_2} = z_{y_2} - \hat{z}_{y_2}$ and the adaptive-gains of the SOSM algorithm $\mu(e_{y_1})$, $\mu(e_{y_2})$ are designed according to (7.31), (7.32). The value of $\dot{z}_{y_1}$ is obtained from the robust exact finite time differentiator (7.19) in [24]. The oxygen excess ratio λ_{O_2} and the fault signal $f(t)$ are estimated

$$\hat{\lambda}_{O_2} = \frac{c_{18}(x_4 - x_1)}{c_{19}\hat{\theta}}, \quad \hat{f}(t) = \frac{\mu(e_{y_2})}{c_5}. \tag{7.56}$$

Remark 7.8 The Assumptions 7.3 and 7.6 are satisfied by the functions $W_{g_1}(y, z_2, u)$, $W_{g_2}(y, z_2, u)$, $\Phi_1(y, u)$, $\Phi_2(y, u)$ and $g_2(y, z_2, u)$. The Riccati equation in Assumption 7.3 is satisfied by appropriate value of the design gain $H = 0.5$. The Assumption 7.4 is also satisfied for some $\epsilon = 0.02$, since the Eq. (7.26) is simplified into a scalar equation.

7.6 Experimental Results

The structure of the observer based fault reconstruction strategy is shown in Fig. 7.2. The nominal values of the parameters for the HIL emulator are shown in Table 4.4. In order to test the robustness of the proposed observer based fault reconstruction

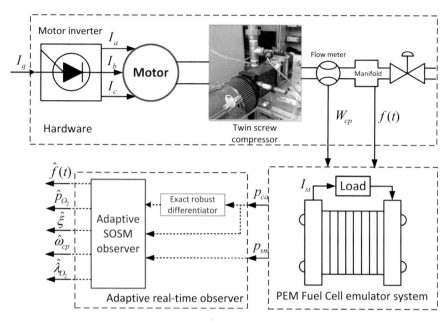

Fig. 7.2 Schematic diagram of the observer based fault reconstruction

approach against parametric uncertainty, the system parameters, i.e. temperatures, volumes, cathode inlet and outlet orifice, motor constant and compressor inertia have been varied around their nominal values according to the worst-case percentages, are given in Table 5.2.

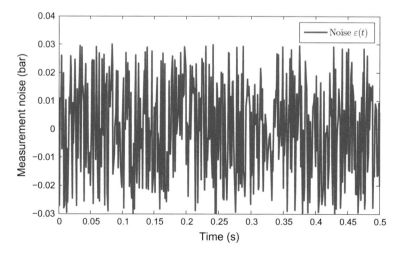

Fig. 7.3 System output measurement noise

During the tests, the stack current was varied between 100 and 450 A, corresponding to flow rate variation in the compressor between 7 and 28 g/s. The measurement

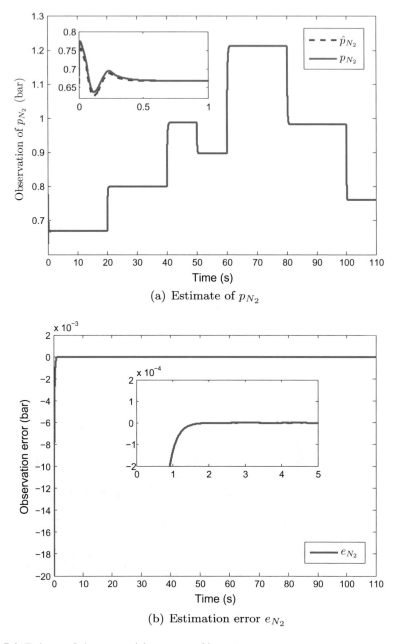

(a) Estimate of p_{N_2}

(b) Estimation error e_{N_2}

Fig. 7.4 Estimate of nitrogen partial pressure and its error

noise was also included to test the robustness of the proposed approach, that is, $y_1 = p_{ca} + \epsilon(t)$, $y_2 = p_{sm} + \epsilon(t)$, where $\epsilon(t)$ represents the measurement noise shown in Fig. 7.3. The state observations (p_{N_2}, ω_{cp}) in response to steps changes in the load current are shown in Figs. 7.4 and 7.5. It can be seen from these figures that

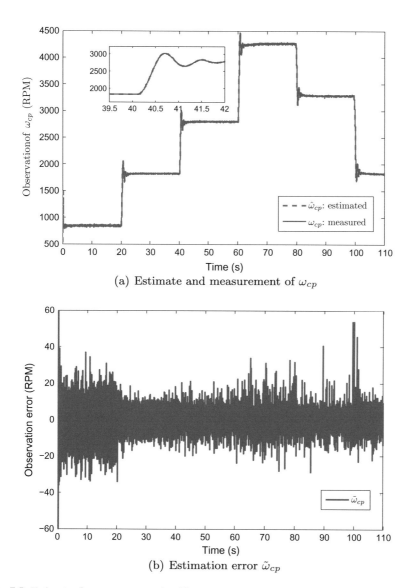

(a) Estimate and measurement of ω_{cp}

(b) Estimation error $\tilde{\omega}_{cp}$

Fig. 7.5 Estimate of compressor speed and its error

the settling time of both upwards load changes (i.e. t = 40 s, t = 60 s) and down-
wards load changes (i.e. t = 80 s, t = 100 s) is less than 2 s. It should be pointed out
that the estimate of compressor speed is compared with experimental measurement.

The outputs of the adaptive differentiator and Levant's fixed gain differentiator
[24] are shown in Fig. 7.6. It is easy to see that the proposed adaptive differentiator
has faster convergence and less overshoot. Figure 7.7 shows the estimate of the stack

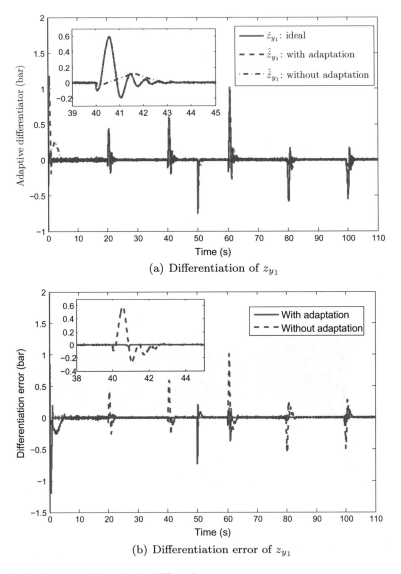

(a) Differentiation of z_{y_1}

(b) Differentiation error of z_{y_1}

Fig. 7.6 Performance of the adaptive differentiator

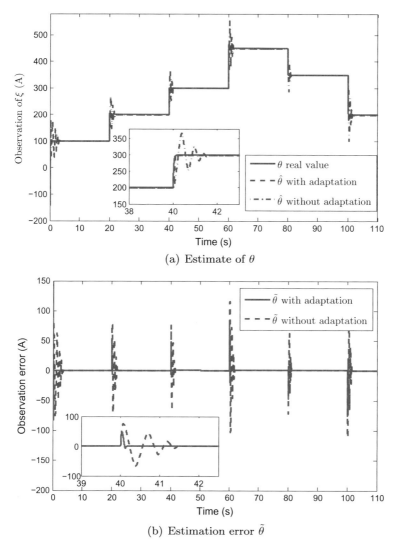

(a) Estimate of θ

(b) Estimation error $\tilde{\theta}$

Fig. 7.7 Estimate of stack current and its error

current, as an unknown parameter θ, based on the adaptive law. Both, the result obtained from application of Levant's differentiator and the adaptive differentiator are presented. It can be seen that the adaptive differentiator improves the performance of the adaptive law and gives a good estimate for the stack current. The fault signal is reconstructed faithfully as shown in Fig. 7.8. It is clear that the proposed scheme is capable of reconstructing fault signal and state estimation simultaneously in the presence of uncertain parameter. The estimate of oxygen excess ratio is shown in Fig. 7.9a. It is easy to see that this value decreases after t = 50 s. This is due to the

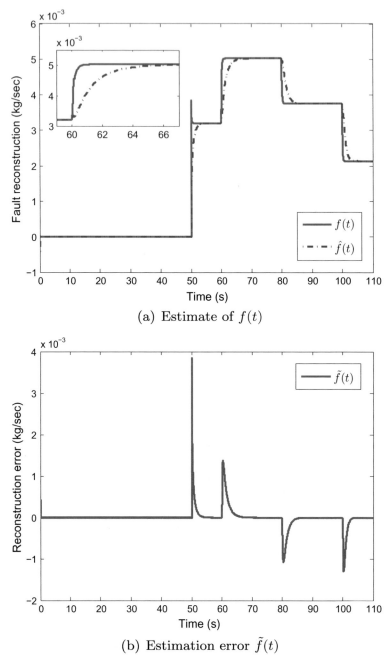

(a) Estimate of $f(t)$

(b) Estimation error $\tilde{f}(t)$

Fig. 7.8 Fault reconstruction and its error

occurrence of the fault (air leak in the supply manifold) after $t = 50\,$s. In this case, the air flow supplied by the compressor needs to increase, taking into account the effect of the fault, in order to ensure safe operation of the fuel cell. The time history of the adaptive gain $L_o(t)$ is shown in Fig. 7.9b, where the exponential convergence

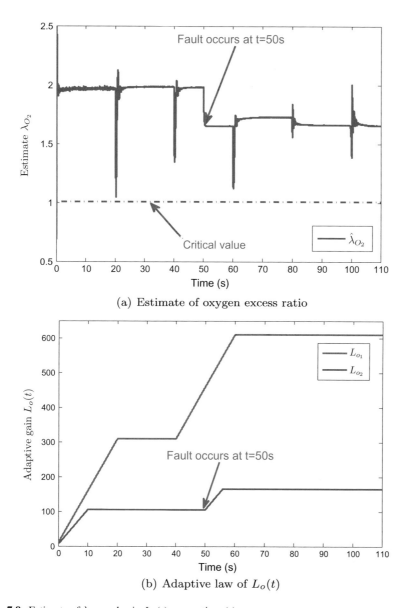

(a) Estimate of oxygen excess ratio

(b) Adaptive law of $L_o(t)$

Fig. 7.9 Estimate of λ_{O_2} and gain $L_o(t)$ versus time (s)

of the observer is ensured. The gains of the SOSM algorithm stop increasing when the output observation error converges to zero in finite time.

7.7 Conclusion

This Chapter has presented a robust fault reconstruction method for a class of nonlinear uncertain systems with Lipschitz nonlinearities based on an adaptive SOSM observer. An adaptive update law has been given to identify the uncertain parameter. The estimated parameter is then injected into an adaptive-gain SOSM observer, which maintains a sliding motion in the presence of the fault signal. The proposed fault reconstruction approach has been successfully applied to the PEMFC system considering a sudden air leak in the supply manifold. The experimental results have shown that the proposed approach is effective and feasible.

References

1. Alwi, H., Edwards, C.: Fault detection and fault-tolerant control of a civil aircraft using a sliding-mode-based scheme. IEEE Trans. Control Syst. Technol. **16**(3), 499–510 (2008)
2. Arcak, M., Görgün, H., Pedersen, L.M., Varigonda, S.: A nonlinear observer design for fuel cell hydrogen estimation. IEEE Trans. Control Syst. Technol. **12**(1), 101–110 (2004)
3. Besançon, G.: High-gain observation with disturbance attenuation and application to robust fault detection. Automatica **39**(6), 1095–1102 (2003)
4. Carrette, L., Friedrich, K., Stimming, U.: Fuel cells-fundamentals and applications. Fuel Cells **1**(1), 5–39 (2001)
5. Chen, J., Patton, R., Zhang, H.: Design of unknown input observers and robust fault detection filters. Int. J. Control **63**(1), 85–105 (1996)
6. Dotelli, G., Ferrero, R., Stampino, P.G., Latorrata, S., Toscani, S.: Diagnosis of pem fuel cell drying and flooding based on power converter ripple. IEEE Trans. Instrum. Meas. **63**(10), 2341–2348 (2014)
7. Edwards, C., Spurgeon, S., Patton, R.: Sliding mode observers for fault detection and isolation. Automatica **36**(4), 541–553 (2000)
8. Edwards, C., Tan, C.P.: Sensor fault tolerant control using sliding mode observers. Control Eng. Pract. **14**(8), 897–908 (2006)
9. Escobet, T., Feroldi, D., Lira, S.D., Puig, V., Quevedo, J., Riera, J., Serra, M.: Model-based fault diagnosis in pem fuel cell systems. J. Power Sources **192**(1), 216–223 (2009)
10. Forrai, A., Funato, H., Yanagita, Y., Kato, Y.: Fuel-cell parameter estimation and diagnostics. IEEE Trans. Energy Convers. **20**(3), 668–675 (2005)
11. Frank, P., Ding, X.: Survey of robust residual generation and evaluation methods in observer-based fault detection systems. J. Process Control **7**(6), 403–424 (1997)
12. Fridman, L., Shtessel, Y., Edwards, C., Yan, X.G.: Higher-order sliding-mode observer for state estimation and input reconstruction in nonlinear systems. Int. J. Robust Nonlinear Control **18**(4–5), 399–412 (2008)
13. Görgün, H., Arcak, M., Barbir, F.: An algorithm for estimation of membrane water content in PEM fuel cells. J. Power Sources **157**(1), 389–394 (2006)
14. Hammouri, H., Kinnaert, M., El Yaagoubi, E.: Observer-based approach to fault detection and isolation for nonlinear systems. IEEE Trans. Autom. Control **44**(10), 1879–1884 (1999)

15. Hou, M., Muller, P.: Design of observers for linear systems with unknown inputs. IEEE Trans. Autom. Control **37**(6), 871–875 (1992)
16. Hou, M., Patton, R.: An LMI approach to H_-/H_∞ fault detection observers. In: UKACC International Conference on Control, UK, pp. 305–310 (1996)
17. Ingimundarson, A., Stefanopoulou, A., McKay, D.: Model-based detection of hydrogen leaks in a fuel cell stack. IEEE Trans. Control Syst. Technol. **16**(5), 1004–1012 (2008)
18. Jiang, B., Staroswiecki, M., Cocquempot, V.: Fault estimation in nonlinear uncertain systems using robust/sliding-mode observers. IEE Proc. Control Theory Appl. **151**(1), 29–37 (2004)
19. Kelouwani, S., Adegnon, K., Agbossou, K., Dube, Y.: Online system identification and adaptive control for PEM fuel cell maximum efficiency tracking. IEEE Trans. Energy Convers. **27**(3), 580–592 (2012)
20. Khalil, H.K.: Nonlinear Systems. Prentice Hall (2001)
21. Kunusch, C., Puleston, P.F., Mayosky, M.A., Fridman, L.: Experimental results applying second order sliding mode control to a PEM fuel cell based system. Control Eng. Pract. **21**(5), 719–726 (2013)
22. Larminie, J., Dicks, A., McDonald, M.S.: Fuel Cell Systems Explained, vol. 2. Wiley, Chichester (2003)
23. Levant, A.: Sliding order and sliding accuracy in sliding mode control. Int. J. Control **58**(6), 1247–1263 (1993)
24. Levant, A.: Robust exact differentiation via sliding mode technique. Automatica **34**(3), 379–384 (1998)
25. Levant, A.: Higher-order sliding modes, differentiation and output-feedback control. Int. J. Control **76**(9–10), 924–941 (2003)
26. Lira, S.D., Puig, V., Quevedo, J., Husar, A.: LPV observer design for PEM fuel cell system: application to fault detection. J. Power Sources **196**(9), 4298–4305 (2011)
27. Liu, J., Laghrouche, S., Harmouche, M., Wack, M.: Adaptive-gain second-order sliding mode observer design for switching power converters. Control Eng. Pract. **30**, 124–131 (2014)
28. Massoumnia, M.A., Verghese, G.C., Willsky, A.S.: Failure detection and identification. IEEE Trans. Autom. Control **34**(3), 316–321 (1989)
29. Matraji, I., Laghrouche, S., Jemei, S., Wack, M.: Robust control of the PEM fuel cell air-feed system via sub-optimal second order sliding mode. Appl. Energy **104**, 945–957 (2013)
30. Moreno, J., Osorio, M.: A Lyapunov approach to second-order sliding mode controllers and observers. In: 47th IEEE Conference on Decision and Control (CDC), pp. 2856–2861. IEEE (2008)
31. Persis, C.D., Isidori, A.: A geometric approach to nonlinear fault detection and isolation. IEEE Trans. Autom. Control **46**(6), 853–865 (2001)
32. Pukrushpan, J., Stefanopoulou, A., Peng, H.: Control of fuel cell breathing: initial results on the oxygen starvation problem. IEEE Control Syst. Mag. **24**(2), 30–46 (2004)
33. Pukrushpan, J.T., Stefanopoulou, A.G., Peng, H.: Control of fuel cell power systems: principles, modeling, analysis and feedback design. Springer Science & Business Media (2004)
34. Qiu, Z., Gertler, J.: Robust FDI systems and H_∞-optimization-disturbances and tall fault case. In: Proceedings of the 32nd IEEE Conference on Decision and Control, USA, pp. 1710–1715 (1993)
35. Rajamani, R.: Observers for Lipschitz nonlinear systems. IEEE Trans. Autom. Control **43**(3), 397–401 (1998)
36. Rakhtala, S.M., Noei, A.R., Ghaderi, R., Usai, E.: Design of finite-time high-order sliding mode state observer: a practical insight to PEM fuel cell system. J. Process Control **24**(1), 203–224 (2014)
37. Ramos-Paja, C., Giral, R., Martinez-Salamero, L., Romano, J., Romero, A., Spagnuolo, G.: A PEM fuel-cell model featuring oxygen-excess-ratio estimation and power-electronics interaction. IEEE Trans. Ind. Electron. **57**(6), 1914–1924 (2010)
38. Romero-Cadaval, E., Spagnuolo, G., Franquelo, L.G., Ramos-Paja, C.A., Suntio, T., Xiao, W.M.: Grid-connected photovoltaic generation plants components and operation. IEEE Ind. Electron. Mag. **7**(3), 6–20 (2013)

39. She, Y., Baran, M., She, X.: Multiobjective control of PEM fuel cell system with improved durability. IEEE Trans. Sustain. Energy **4**(1), 127–135 (2013)
40. Slotine, J.J.E., Li, W., et al.: Applied Nonlinear Control, vol. 199. Prentice Hall, New Jersey (1991)
41. Spurgeon, S.K.: Sliding mode observers: a survey. Int. J. Syst. Sci. **39**(8), 751–764 (2008)
42. Suh, K.W., Stefanopoulou, A.G.: Performance limitations of air flow control in power-autonomous fuel cell systems. IEEE Trans. Control Syst. Technol. **15**(3), 465–473 (2007)
43. Talj, R.J., Hissel, D., Ortega, R., Becherif, M., Hilairet, M.: Experimental validation of a PEM fuel-cell reduced-order model and a moto-compressor higher order sliding-mode control. IEEE Trans. Ind. Electron. **57**(6), 1906–1913 (2010)
44. Tan, C., Edwards, C.: Sliding mode observers for detection and reconstruction of sensor faults. Automatica **38**(10), 1815–1821 (2002)
45. Tan, C., Edwards, C.: Sliding mode observers for robust detection and reconstruction of actuator and sensor faults. Int. J. Robust Nonlinear Control **13**(5), 443–463 (2003)
46. Vahidi, A., Stefanopoulou, A., Peng, H.: Current management in a hybrid fuel cell power system: a model-predictive control approach. IEEE Trans. Control Syst. Technol. **14**(6), 1047–1057 (2006)
47. Vazquez, S., Sanchez, J.A., Carrasco, J.M., Leon, J.I., Galvan, E.: A model-based direct power control for three-phase power converters. IEEE Trans. Ind. Electron. **55**(4), 1647–1657 (2008)
48. Veluvolu, K.C., Defoort, M., Soh, Y.C.: High-gain observer with sliding mode for nonlinear state estimation and fault reconstruction. J. Franklin Inst. **351**(4), 1995–2014 (2014)
49. Wang, H., Huang, Z.J., Daley, S.: On the use of adaptive updating rules for actuator and sensor fault diagnosis. Automatica **33**, 217–225 (1997)
50. Yan, X., Edwards, C.: Nonlinear robust fault reconstruction and estimation using a sliding mode observer. Automatica **43**(9), 1605–1614 (2007)
51. Yan, X.G., Edwards, C.: Adaptive sliding-mode-observer-based fault reconstruction for nonlinear systems with parametric uncertainties. IEEE Trans. Ind. Electron. **55**(11), 4029–4036 (2008)
52. Yang, H., Saif, M.: Fault detection in a class of nonlinear systems via adaptive sliding observer. IEEE Int. Conf. Syst. Man Cybern. **3**, 2199–2204 (1995)
53. Yang, H., Saif, M.: Nonlinear adaptive observer design for fault detection. In: Proceedings of the American Control Conference, USA, pp. 1136–1139 (1995)
54. Zhang, X., Polycarpou, M., Parisini, T.: Fault diagnosis of a class of nonlinear uncertain systems with lipschitz nonlinearities using adaptive estimation. Automatica **46**(2), 290–299 (2010)
55. Zhang, Y., Jiang, J.: Bibliographical review on reconfigurable fault-tolerant control systems. Annu. Rev. Control **32**(2), 229–252 (2008)
56. Ziogou, C., Papadopoulou, S., Georgiadis, M.C., Voutetakis, S.: On-line nonlinear model predictive control of a PEM fuel cell system. J. Process Control **23**(4), 483–492 (2013)

Chapter 8
Sliding Mode Control of DC/DC Power Converters

In the previous chapters, we have successfully designed adaptive HOSM based observers for state observation and FDI of the PEM fuel cell systems. We now turn our attention towards the power side of the PEMFC. In fact, the PEMFC itself has severe dynamic limitations due to the time response of fuel flow and fuel delivery systems (hydrogen and air feed systems) [23]. In order for a PEMFC power system to be employed in varying load applications like electrical vehicles, storage elements with fast response time need to be integrated into the system. A typical PEMFC power system usually relies on rechargeable batteries, super-capacitors or both for improved power dynamic characteristics in transient high-power demands.

Evidently, such a hybrid power system requires power conditioning circuits with precise power control algorithms behind them such that the output power is compatible with the constraints of the power bus. These circuits include DC/DC converters which are used to control the voltages of the fuel cell and storage elements to a desired value, rectifiers and inverters that are used to convert AC to DC and vice versa. Therefore, precise control, and therefore observation, are equally important in this area of the PEMFC system as they were in the air-feed system.

In this chapter, we have studied the control and observation of DC/DC power converters that are employed in fuel cell hybrid power systems, using sliding mode techniques. The main focus is on the necessary modification and improvement of conventional sliding mode methods for their applications to fuel cell power systems.

8.1 Introduction

As a ubiquitous direct current (DC) voltage conversion technology in industrial applications, DC/DC converters have been widely applied in the uninterruptible power supply, telecommunication equipment, DC motor drives, power grids, and clean

© Springer Nature Switzerland AG 2020
J. Liu et al., *Sliding Mode Control Methodology in the Applications of Industrial Power Systems*, Studies in Systems, Decision and Control 249,
https://doi.org/10.1007/978-3-030-30655-7_8

energy systems [7]. DC/DC converters generally include boost type [26], buck type [31], and buck-boost type [18] converters. In DC/DC boost converters, one DC voltage can be converted to another higher DC voltage through a converter [2]. Nowadays, this converter is widely used in remote and data communications, computers, office automation equipments, industrial instruments, military, aerospace and other fields [3, 11]. For an accurate voltage regulation of the DC/DC boost converters, the output voltage of the converters to be controlled is always a challenging problem due to the existence of various uncertainties.

To represent the nonlinear systems with complex uncertainties, the interval type-2 (IT2) fuzzy models [10] were proposed, which are established by applying type-2 fuzzy sets. They allow uncertain parameters or unknown variables to exist in the membership functions of the fuzzy models. This modeling methodology has substantially extended its utilization on the uncertain systems, and it has also been demonstrated that the IT2 fuzzy model is less conservative than the type-1 fuzzy model based on type-1 fuzzy sets [9]. Nevertheless, the defect of this method is the lack of self-learning and self-organizing ability. What's more, fuzzy neural networks (FNNs) [1] can make up for the deficiency, and many online learning algorithms have been studied for the updating of the IT2FNNs [6, 12, 13]. To mention a few, the authors in [6] developed an IT2FNN with support-vector regression to deal with noisy regression problems. [13] simplified a type of self-evolving IT2FNN for reducing the time-consuming K-M iterative steps, by using a gradient descent algorithm. For the problems of noise cancellation and system modeling, a Takagi-Sugeno-Kang-type-based self-evolving compensatory IT2FNN was designed in [12]. This approach combines the ability of IT2 fuzzy reasoning to manage uncertainties and the capability of artificial FNNs to learn from processes. Hence, one of the motivations of this work is the utilization of the IT2FNNs to the representation of a more practical boost converter model with uncertain parameters.

For the control of the DC/DC converters, traditional proportional-integral-type or proportional-integral-derivative-type controllers are the common control method [19] in practice. But the existence of the uncertainties in the DC/DC converters remains the difficulty to the stability analysis of the corresponding tracking error systems with such a control. Recently, intelligent control methods like fuzzy control, adaptive control, backstepping technique control and SMC, have attracted considerable research interests in dealing with the control problem of the DC/DC boost converters. What needs to be emphasized is that, among these control methods, SMC technique [15, 16, 25, 30, 32] has been recognized as a very effective and popular method due to its capability to achieve wide stability range, fast dynamic response, strong robustness, and simple control design procedure. Benefited from the SMC strategy, significant robust control methods have been developed for the voltage regulation of the DC/DC boost converters over the past few decades. For example, a robust adaptive sliding mode controller was designed in [20] using state observers for the boost converter with an external input voltage load and unknown resistive. The work in [21] presented an adaptive SMC approach to the DC/DC converters, of which the state-space model was formulated with some unknown coordinate components of the system equilibrium. The authors in [26] designed an SMC scheme for the robust voltage control of the DC/DC converters. Then, by using the FNN, which is a type-1 FNN with Gauss-type membership functions, an integral SMC law was designed in [27] for the DC/DC converters. However, it requires a complex calculation of the adaptive parameters of the FNN for the updating of the SMC law, and the alleviated chattering actually still exists in the designed SMC effort due to its discontinuity.

Inspired by the above discussion, in this paper, we will design a continuous SMC law, that is, a *dynamic SMC*, for the DC/DC boost converters, by utilizing a simplified IT2FNN. In terms of the representation of the uncertainties, we will fully formulate the parameter perturbation of input inductor, output capacitor, load resistor, input voltages, and noises in the DC/DC converters. The unknown part of the formulated uncertainties is estimated by a simplified ellipsoidal-type IT2FNN. Then, the dynamic SMC law will be designed based on both artificial FNN and adaptive IT2FNN. The adaptive rules of the IT2FNN are designed using the gradient descent method (GDM). Comparative simulations are detailed to illustrate the effectiveness of the designed control scheme.

8.2 Modeling of Boost Type DC/DC Converters

A typical boost DC/DC converter is considered, which transforms the DC fuel cell stack power to the external power devices with possible voltage-current requirements. Boost converters are used for applications in which the required output voltage is higher than the source voltage. From a control point of view, the boost converters are difficult because its standard model is a non-minimum phase system. Traditionally, the control problems of the DC/DC converters are solved by using pulse-width modulation (PWM) techniques. Sira-Ramirez [22] demonstrated the equivalence between SMC and PWM control in the low frequency range for a boost converter. In view of practical implementation, SMC is much easier than a PWM control since the maximum frequency of commercially available switching elements increases higher and higher [24]. The sliding mode approach is expected to become increasingly popular in the field of power converter control.

Figure 8.1 illustrates the circuit topology of a common boost converter, where D is the output diode, S is the main switch, and V_i, L, R and C are input voltage, input inductor, the load resistor and output capacitor, respectively. u denotes the duty cycle, which is the percentage of the pulse width d over the total period τ between an ON-and-OFF cycle of the switch. This is to say $u = \frac{d}{\tau}$ for some predefined pulse width. The switching ON or OFF is determined by the PWM-based duty cycle. Generally, the parameters V_i, L, R and C are usually treated as ideal constants. But, in practice, these parameters should be treated as variables due to the existence of measurement errors or changes in actual application requirements. By applying the state-space averaging method to the DC/DC boost converter, the average model is formulated as

Fig. 8.1 Topology of the DC/DC boost converter: **a** Switching OFF state, and **b** Switching ON state

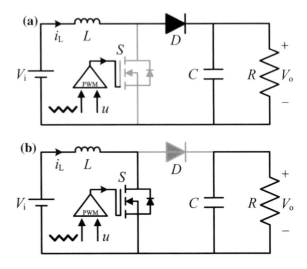

$$i_L = -\frac{(1-u)}{L} V_o + \frac{1}{L} V_i, \tag{8.1}$$

$$\dot{V}_o = \frac{(1-u)}{C} i_L - \frac{1}{RC} V_o + \frac{i_{ld}}{C}, \tag{8.2}$$

where V_o denotes the average output (capacitor) voltage, i_L stands for the average inductor current, $u \in (0, 1)$ is also called the average control effort here, and the considered i_{ld} denotes the current source, which is caused by unmodeled dynamics, load variations, and unpredictable uncertainties.

Further, considering the occurrence of the uncertainties caused by parameter perturbations, a practical average model of the studied converter is further formulated as

$$i_L = -\frac{(1-u)}{L+\Delta L} V_o + \frac{1}{L+\Delta L} (V_i + \Delta V_i), \tag{8.3}$$

$$\dot{V}_o = \frac{(1-u)}{C+\Delta C} i_L - \frac{1}{(R+\Delta R)(C+\Delta C)} V_o + \frac{i_{ld}}{C+\Delta C}, \tag{8.4}$$

where ΔL, ΔC, ΔR and ΔV_i are the perturbations of L, C, R and V_i, respectively.

8.2.1 Representation of the Tracking Error System with Uncertain Parameters

Define the following tracking error variables:

$$e_i = i_{ref} - i_L, \tag{8.5}$$

$$e_V = V_{ref} - V_o, \tag{8.6}$$

where i_{ref} is a reference input of the inductor current and V_{ref} is the command of the desired output voltage. The dynamics of the tracking error system is thus derived as

$$\dot{e} = Ae + bu + Q + W, \tag{8.7}$$

where

$$e = \begin{bmatrix} e_i & e_V \end{bmatrix}^{\mathrm{T}}, \ b = \tilde{B}x, \ x(t) = \begin{bmatrix} i_L & V_o \end{bmatrix}^{\mathrm{T}},$$

$$A = \begin{bmatrix} 0 & -\frac{1}{L} \\ \frac{1}{C} & 0 \end{bmatrix}, \ \tilde{B} = \begin{bmatrix} 0 & \frac{1}{L+\Delta L} \\ -\frac{1}{C+\Delta C} & 0 \end{bmatrix}, \ Q = \begin{bmatrix} q_1(t) \\ q_2(t) \end{bmatrix}, \ W = \begin{bmatrix} w_1(t) \\ w_2(t) \end{bmatrix},$$

$$q_1(t) = \frac{L\Delta V_i - (V_i - V_o)\,\Delta L}{L\,(L+\Delta L)},$$

$$q_2(t) = \frac{(R\Delta C + C\Delta R + \Delta R\Delta C)\,V_o}{RC\,(RC + R\Delta C + C\Delta R + \Delta R\Delta C)} - \frac{i_L\Delta C}{C\,(C+\Delta C)} + \frac{i_{ld}}{C+\Delta C},$$

$$w_1(t) = \frac{V_{ref} - V_i}{L} + \dot{i}_{ref}, \ w_2(t) = \frac{V_o}{RC} - \frac{i_{ref}}{C} + \dot{V}_{ref}.$$

In this paper, we assume the allowed parameter perturbations satisfy $-aL \le \Delta L \le aL$, $-bC \le \Delta C \le bC$, and $-cR \le \Delta R \le cR$, where a, b and c are prior known constants belonging to $(0, 1)$. Considering the uncertainties in $\tilde{B}$, we represent the uncertainty terms by the utilization of the fuzzy model approximation [4, 14] as follows:

$$\tilde{B} = \hat{B} + \Delta B = \sum_{i=1}^{2}\sum_{j=1}^{2} h_{Li}(\Delta \hat{L})h_{Cj}(\Delta \hat{C})(B_{Li} + B_{Cj}) + \Delta B, \tag{8.8}$$

where $-aL \le \Delta \hat{L} = aL\text{rand} \le aL$ and $-bL \le \Delta \hat{C} = bC\text{rand} \le bL$ with $0 < a < 1$ and $0 < b < 1$ denote the estimates of ΔL and ΔC, respectively, used in this paper, $\hat{B}$ is the approximated matrix variable that will be used in further design procedure, ΔB is the approximation error, and

$$B_{Li} = \begin{bmatrix} 0 & f_{1\min} \\ 0 & 0 \end{bmatrix}, \qquad\qquad B_{Li} = \begin{bmatrix} 0 & f_{1\max} \\ 0 & 0 \end{bmatrix},$$

$$B_{Ci} = \begin{bmatrix} 0 & 0 \\ -f_{2\min} & 0 \end{bmatrix}, \qquad\qquad B_{Ci} = \begin{bmatrix} 0 & 0 \\ -f_{2\max} & 0 \end{bmatrix},$$

$$f_1 = \frac{1}{L+\Delta L}, \qquad\qquad f_2 = \frac{1}{C+\Delta C},$$

$$f_{1\min} = \frac{1}{(1+a)L}, \qquad\qquad f_{1\max} = \frac{1}{(1-a)L},$$

$$f_{2\min} = \frac{1}{(1+b)C}, \qquad\qquad f_{2\max} = \frac{1}{(1-b)C},$$

$$h_{L1} = \frac{f_{1\,\text{max}} - f_1}{f_{1\,\text{max}} - f_{1\,\text{min}}}, \qquad\qquad h_{L2} = 1 - h_{L1},$$

$$h_{C1} = \frac{f_{2\,\text{max}} - f_2}{f_{2\,\text{max}} - f_{2\,\text{min}}}, \qquad\qquad h_{C2} = 1 - h_{C1}.$$

Remark 8.1 It should be noted that these two membership functions are not the component of the fuzzy engine, but only the approximation or estimation of $\tilde{B}$ by using $\Delta\hat{L}$ and $\Delta\hat{C}$ chosen in this paper. Suppose the approximation error ΔB bounded by $||\Delta B|| \leq \beta$. β can be evaluated by repeated simulations, and it has been validated in the simulation that $||\Delta B|| < 0.5||\hat{B}||$ no matter how the interval uncertainties ΔL and ΔC change. Based on the approximation (8.8), the error dynamics (8.7) can be rewritten as

$$\dot{e} = Ae + (\hat{B} + \Delta B)xu + Q + W. \tag{8.9}$$

8.2.2   A Simplified Ellipsoidal-Type IT2FNN

In order to fully formulate the uncertainties existing in the system dynamics, we employ an IT2FNN model to represent the tracking error system (8.7). For some input variable $z = [z_1\ z_2\ z_3]^{\text{T}} = [i_L\ V_o\ u]^{\text{T}}$, the IF-THEN rules are presented as follows:

◊ Fuzzy Rule i: IF z_1 is S_{1i}, and z_2 is S_{2i} and z_3 is S_{3i}, THEN q is $[q_i^L, q_i^U]$, where S_{1i} is an IT2 fuzzy set for the antecedent part, q is the consequent variable represented by a weighting interval set $[q_i^L, q_i^U]$, and $i = 1, 2, \ldots, r$ is the number of the fuzzy rules. Based on the fuzzy rules above, the ellipsoidal-type IT2FNN is detailed as follows.

▷ Input layer: The j-th input is z_j of the input layer for the IT2FNN ($j = 1, 2, 3$).

▷ Membership function layer: For the j-th input and i-th node of the member-ship layer, we adopt the modified ellipsoidal-type of lower and upper membership functions [8], which is a kind of exponential-type type-2 membership functions with simplified computation complexity and more robustness [29], as follows:

$$\underline{g}_{ij}(z_j) = \begin{cases} 1 - \left|\dfrac{x_j - m_{ij}}{v_{ij}}\right|^{\alpha}, & \text{if } |x_j - m_{ij}| < v_{ij}, \\ 0, & \text{else,} \end{cases} \tag{8.10}$$

$$\bar{g}_{ij}(z_j) = \begin{cases} 1 - \left|\dfrac{x_j - m_{ij}}{v_{ij}}\right|^{\frac{1}{\alpha}}, & \text{if } |x_j - m_{ij}| < v_{ij}, \\ 0, & \text{else,} \end{cases} \tag{8.11}$$

where $0 < \alpha < 1$, m_{ij} and v_{ij} denote respectively the center and the width of the membership function. α, chosen as $0 < \alpha < 1$ in this paper, is the parameter to adjust the shape of the footprint of uncertainty (FOU) of the exponential-type type-2 membership functions.

▷ Rule layer: Based on the lower and upper membership functions in Membership function layer, the i-th node in this layer is the product result for the interval degree of firing $\left[\underline{h}_i, \bar{h}_i\right]$ calculated by

$$\underline{h}_i = \prod_{j=1}^{3} \underline{g}_{ij}(z_j), \quad \bar{h}_i = \prod_{j=1}^{3} \bar{g}_{ij}(z_j). \tag{8.12}$$

▷ Type-reduction layer: This layer plays a role of reducing the type-2 fuzzy set to a type-reduced one for a crisp output [17]. Considering the consequent interval weighting $q = [q_i^L, q_i^U]$, we assign the bounds q_i^L and q_i^U in ascending order, i.e., $q_1^L \le q_2^L \le \cdots \le q_r^L$ and $q_1^U \le q_2^U \le \cdots \le q_r^U$. Considering the computation complexity, we use a simplified type of reduction expressed by

$$m_L = \frac{\sum_{i=1}^{r} \underline{h}_i q_i^L}{\sum_{l=1}^{r} \underline{h}_l} \triangleq q_L^T \underline{h}, \quad m_U = \frac{\sum_{i=1}^{r} \bar{h}_i q_i^U}{\sum_{l=1}^{r} \bar{h}_l} \triangleq q_U^T \bar{h},$$

where

$$q_L \triangleq \left[q_1^L, q_2^L, \ldots, q_r^L\right]^T \in \mathbb{R}^r, \quad q_U \triangleq \left[q_1^U, q_2^U, \ldots, q_r^U\right]^T \in \mathbb{R}^r,$$

$$\underline{h} \triangleq \left[h_1^L, q_2^L, \ldots, h_r^L\right]^T \in \mathbb{R}^r, \quad h_i^L \triangleq \frac{\underline{h}_i}{\sum_{l=1}^{r} \underline{h}_l},$$

$$\bar{h} \triangleq \left[h_1^U, h_2^U, \ldots, h_r^U\right]^T \in \mathbb{R}^r, \quad h_i^U \triangleq \frac{\bar{h}_i}{\sum_{l=1}^{r} \bar{h}_l}.$$

Further, in order to guarantee $h_i^L \le h_i^U$, we update $h_i^L := \min\left\{h_i^L, h_i^U\right\}$ and $h_i^U := \max\left\{h_i^L, h_i^U\right\}$. $\underline{h}$ and $\bar{h}$ are the firing strength vectors. To avoid singular values in actual computation, we here set $h_i^L = 0$ if $\sum_{l=1}^{r} \underline{h}_l = 0$, and $h_i^U = 0$ if $\sum_{l=1}^{r} \bar{h}_l = 0$.

▷ Output layer: we take the output of the IT2FNN from the average of m_L and m_U:

$$q = \frac{m_L + m_U}{2}. \tag{8.13}$$

8.2.3 Control Objective

For the voltage regulation of the considered boost converter, in the following, we aim to firstly design a dynamic SMC against the uncertainties existing in the DC/DC model, based on the represented tracking error dynamics (8.9). For a given command of the desired output voltage V_{ref}, the capacitor voltage V_o will accurately track V_{ref} equivalently to minimize the tracking error of the output voltage by properly manipulating the switch, which is determined by some designed duty cycle input u. This is to be achieved despite variations in the considered model system. Meanwhile, the output voltage is to be controlled to tracking the command as fast and with as little overshoot as possible during transients. Furthermore, the uncertainties, including the variations of the load resistance and input voltage, will be eliminated by using the introduced IT2FNN, and then an IT2FNN-based SMC law will be designed for the compensation of the uncertainties, as well as the robust tracking control of the desired inductance current and capacitor voltage of the DC/DC converter.

8.3   Control Design and Stability Analysis

8.3.1   Dynamic SMC Design

8.3.1.1   Establishment of the Sliding Surface

Recalling the fact $0 \leq u \leq 1$ and the relation $V_i = V_o(1 - u)$, in terms of the dynamics (8.9), we propose the following linear sliding surface:

$$s = ge + h(u - v_0) = ge + hv \tag{8.14}$$

where

$$v = u - \left(1 - \frac{V_{\text{in}}}{V_{\text{ref}}}\right) = u - v_0 \in [-1, 1],$$

and $v_0 = 1 - \frac{V_{\text{in}}}{V_{\text{ref}}} \in [0, 1]$, $g = [\, g_1 \ g_2 \,] \in \mathbb{R}^{1 \times 2}$ is a row vector, and $h \in \mathbb{R}$ is a nonzero constant. Further, we consider $\dot{V}_i = \rho$ and $\dot{V}_{\text{ref}} = \sigma$ as some known constants. Considering the sliding motion of s, that is $\dot{s} = s = 0$, we obtain the following dynamic equivalent control law:

$$\dot{v}_{\text{eq}} = -h^{-1}g\left[Ae + (\hat{B} + \Delta B)xu + Q + W\right]. \tag{8.15}$$

Remark 8.2 The sliding surface (8.14) is designed based on the desired final equilibrium determined by the desired voltage and the nominal voltage. Besides the tracking error variable e, the introduced auxiliary variable v is also to be designed to approach to zero. Then, the accurate voltage regulation can be achieved while the sliding surface $s = 0$ is reached.

Due to the existence of the uncertainties in $\dot{v}_{\text{eq}}$ and the requirement of fast response, we introduce the following switching law:

$$\dot{v}_s = -\eta \text{sign}(s) - \varsigma s - \zeta|s|^\tau \text{sign}(s), \tag{8.16}$$

where $\eta > 0$ is a gain to be determined, $\varsigma > 0$, and $\zeta > 0$ and $0 < \tau < 1$ are predefined constants. v_s is actually an exponential rate plus power rate reaching law [5], which is designed to properly and effectively reduce the reaching time of sliding mode and weaken the chattering caused by the switching term.

Hence, based on $\dot{v}_{eq}$ and (8.16), we propose firstly the following dynamic SMC law:

$$\dot{v} = -h^{-1}g(Ae + \hat{B}xv + \hat{B}xv_0 + W) - \eta\,\text{sign}(s) - \varsigma s - \zeta|s|^\tau\text{sign}(s), \quad (8.17)$$

with $\eta = \gamma\|x\|$, $\varsigma > 0$, and $\zeta > 0$ and $0 < \tau < 1$ considering $\|\Delta B\| \cdot \|x\| \cdot \|u\| + \|Q\| \leq \gamma\|x\|$, and γ can be calculated with the interval uncertainties, and thus we omit the expression of γ here. According to the proposed SMC law for the tracking error system (8.9), let us analyze the reachability of the desired sliding mode in the following theorem.

Theorem 8.1 *Consider the dynamics (8.9) and (8.17), and the sliding surface (8.14). For some given initial conditions $u(0)$ and $e(0)$, and ς, ζ and τ in (8.17), the trajectory of e can be forced onto the sliding surface $s = 0$ in finite time by the dynamic SMC law.*

Proof Let us check the reachability condition $s\dot{s} < 0$, $\forall s \neq 0$, using the SMC law (8.17). For all $s \neq 0$, from (8.9), (8.14) and (8.17), it follows that

$$\begin{aligned}
s\dot{s} &= s(g\dot{e} + h\dot{v}) \\
&= sg\Big[Ae + (\hat{B} + \Delta B)xv + (\hat{B} + \Delta B)xv_0 + Q + W\Big] \\
&\quad -sg(Ae + \hat{B}xv + \hat{B}xv_0 + W) - \eta s \cdot \text{sign}(s) - \varsigma s^2 - \zeta s \cdot |s|^\tau\text{sign}(s) \\
&\leq -\varsigma s^2 - \zeta|s|^{1+\tau} < 0,
\end{aligned}$$

for $s \neq 0$, where $\varsigma > 0$, $\zeta > 0$ and $\tau > 0$. Hence, the desired sliding mode is reachable and can be achieved in finite time under the control law (8.17). This completes the proof. ∎

When the sliding surface $s = 0$ is reached, the following sliding motion dynamics of e and v can be derived from (8.9) and (8.17).

$$S : \begin{cases} \dot{e} = (A - v_0\hat{B})e - \hat{B}ev + (\Delta Bxu + Q) + \hat{B}x_{\text{ref}}(v + v_0) + W, \\ \dot{v} = -h^{-1}g\Big[(A - v_0\hat{B})e - \hat{B}ev + \hat{B}x_{\text{ref}}(v + v_0) + W\Big]. \end{cases}$$

8.3.1.2 The Existence of the Sliding Motion

On the basis of the resulting dynamic system S, the following theorem states the existence of such a sliding motion S.

Theorem 8.2 *Consider the dynamic system S. For some given nonzero constant h and row vector g, if there exists matrix $0 < P = P^T \in \mathbb{R}^{2\times2}$, $0 < \Gamma = \Gamma^T \in \mathbb{R}^{2\times2}$, and scalar $\delta > 0$, such that for $i, j = 1, 2$,*

$$P_\delta\big[A - v_0(B_{Li} + B_{Cj})\big] + \big[A - v_0(B_{Li} + B_{Cj})\big]^\mathsf{T} P_\delta + \Gamma < 0, \qquad (8.18)$$

where $P_\delta \triangleq P + \delta h^{-2} g^\mathsf{T} g$, then system $\mathcal{S}$ is boundedly stable.

Proof Firstly, from condition (8.18), recalling $h_{L1} + h_{L2} = 1$ and $h_{C1} + h_{C2} = 1$, one can obtain that

$$0 < \Gamma < M \triangleq P_\delta(A - v_0\hat{B}) + (A - v_0\hat{B})^\mathsf{T} P_\delta.$$

Since $\dot{s} = s = 0$ in the sliding stage, we obtain

$$v = -h^{-1} g e,$$

from (8.14). Then, the system $\mathcal{S}$ can be further expressed by

$$\mathcal{S}' : \begin{cases} \dot{e} = (A - v_0\hat{B})e + \big[(\hat{B} + \Delta B)xv + \Delta Bxv_0 + Q\big] + \big(\hat{B}x_{\mathrm{ref}}v_0 + W\big), \\ \dot{v} = -h^{-1} g\big[(A - v_0\hat{B})e + \hat{B}xv + \big(\hat{B}x_{\mathrm{ref}}v_0 + W\big)\big]. \end{cases}$$

In terms of the dynamic system $\mathcal{S}$, we choose the Lyapunov function candidate $V = e^\mathsf{T} P e + \delta v^2$ with matrix $P > 0$ and scalar $\delta > 0$. From the expressed $\dot{e}$ and $\dot{v}$ in $\mathcal{S}'$, it follows that

$$
\begin{aligned}
\dot{V} &= 2e^\mathsf{T} P\dot{e} + 2v\delta\dot{v} \\
&= 2e^\mathsf{T} P\Big\{(A - v_0\hat{B})e + \big[(\hat{B} + \Delta B)xv + \Delta Bxv_0 + Q\big] + (\hat{B}x_{\mathrm{ref}}v_0 + W)\Big\} \\
&\quad -2e^\mathsf{T} g^\mathsf{T} h^{-1}\delta\Big\{-h^{-1} g\big[(A - v_0\hat{B})e + \hat{B}xv + \big(v_0\hat{B}x_{\mathrm{ref}} + W\big)\big]\Big\} \\
&= 2e^\mathsf{T} P(A - v_0\hat{B})e + 2e^\mathsf{T} P\Big\{\big[(\hat{B} + \Delta B)xv \\
&\quad + \Delta Bxv_0 + Q\big] + \big(\hat{B}x_{\mathrm{ref}}v_0 + W\big)\Big\} + 2e^\mathsf{T} qh^{-2} g^\mathsf{T} g(A - v_0\hat{B})e \\
&\quad +2e^\mathsf{T} \cdot qh^{-2} g^\mathsf{T} g\Big[\hat{B}xv + \big(v_0\hat{B}x_{\mathrm{ref}} + W\big)\Big] \\
&= 2e^\mathsf{T}(P + qh^{-2} g^\mathsf{T} g)(A - v_0\hat{B})e + 2e^\mathsf{T}\Big\{P\big[(\hat{B} + \Delta B)xv + Q\big] \\
&\quad +qh^{-2} g^\mathsf{T} g\hat{B}xv + (P + qh^{-2} g^\mathsf{T} g)\big(v_0\hat{B}x_{\mathrm{ref}} + W\big)\Big\} \\
&= 2e^\mathsf{T}(P + qh^{-2} g^\mathsf{T} g)(A - v_0\hat{B})e + 2e^\mathsf{T}\Big\{P\big[(\hat{B} + \Delta B)xv + \Delta Bxv_0 + Q \\
&\quad +qh^{-2} P^{-1} vg^\mathsf{T} g\hat{B}x\big] + (P + qh^{-2} g^\mathsf{T} g)\big(v_0\hat{B}x_{\mathrm{ref}} + W\big)\Big\} \\
&= 2e^\mathsf{T} Me + 2e^\mathsf{T}\big[Pq + (P + qh^{-2} g^\mathsf{T} g)\,w\big] \\
&\le -\lambda_{\min}(\Gamma) \cdot \|e\|^2 + 2\|e\|\big(\|P\| \cdot \|q\| + \|P + qh^{-2} g^\mathsf{T} g\| \cdot \|w\|\big) \\
&\le -\lambda_{\min}(\Gamma)\|e\|\big(\|e\| - r\big),
\end{aligned}
$$

from which one can conclude that $\dot{V} < 0$ for $e \in \mathbb{B}^c(r)$, where $\mathbb{B}^c(r)$ is the complement of the closed ball $\mathbb{B}^c(r)$ centered at $e = 0$ with radius $r = \frac{2(\|P\|\cdot\|q\|+\|(P+qh^{-2}g^{\mathrm{T}}g)\|\cdot\|w\|)}{\lambda_{\min}(\Gamma)}$, where

$$q = (\hat{B} + \Delta B)xv + \Delta Bxv_0 + Q + qh^{-2}P^{-1}g^{\mathrm{T}}g\hat{B}xv,$$
$$w = v_0\hat{B}x_{\mathrm{ref}} + W.$$

Obviously, r is bounded, since the uncertainty in q is bounded and w is known. Meanwhile, r can be minimized by increasing Γ and h, and decreasing δ and g. Hence, the sliding motion (8.14) exists based on condition (8.18) and is boundedly stable. It completes the proof. $\blacksquare$

8.3.2 Ellipsoidal-Type IT2FNN-Based Dynamic SMC Design

In this subsection, we consider the uncertainties of which the bounds are unknown by the representation of the introduced ellipsoidal-type IT2FNN. Define $q = \begin{bmatrix} q_1 & q_2 \end{bmatrix}^{\mathrm{T}}$. Then, based on the considered IT2FNN, q_1 and q_2 represented by $q_k = \frac{1}{2}q_k^{\mathrm{T}}h_k + \delta_k \in \mathbb{R}$ denote the unknown parts in voltage and current dynamics, including the noises, where $q_k = \begin{bmatrix} q_{Lk}^{\mathrm{T}}, q_{Uk}^{\mathrm{T}} \end{bmatrix}^{\mathrm{T}} \in \mathbb{R}^{2r}$, $h_k \triangleq \begin{bmatrix} \underline{h}_k^{\mathrm{T}}, \bar{h}_k^{\mathrm{T}} \end{bmatrix}^{\mathrm{T}} \in \mathbb{R}^{2r}$, and δ_k denotes the approximation error. Thus, in the following, we will design some proper adaptive laws to estimate the unknown parameters q_k. Denote the estimation error $\tilde{q}_k = q_k - \hat{q}_k$. Thus, the dynamics in (8.9) can be further written as

$$\dot{e} = (A - v_0\hat{B})e + \hat{B}ev + (\Delta Bxu + Q) + \hat{B}x_{\mathrm{ref}}(v + v_0) + W$$
$$= (A - v_0\hat{B})e + \hat{B}ev + q(x, u) + \hat{B}x_{\mathrm{ref}}u + W, \tag{8.19}$$

where $q(x, u) = \Delta Bxu + Q$. According to this concept, we modify the dynamic SMC law from (8.17) in the following theorem, for the reachability of the desired sliding mode.

Theorem 8.3 *Consider the dynamic system (8.19) and the IT2FNN in Subsection II-B. The trajectory of e and v can be forced onto the sliding surface $s = 0$ in finite time by the following IT2FNN-based dynamic SMC law:*

$$\dot{v} = -h^{-1}g(Ae + \hat{B}xv + \hat{B}xv_0 + W) - \frac{1}{2}\sum_{k=1}^{2}g_k\hat{q}_k^{\mathrm{T}}h_k - \varsigma s - \zeta|s|^{\tau}\mathrm{sign}(s),$$
$$\tag{8.20}$$

with adaptive laws

$$\dot{\hat{q}}_k = -\kappa_k E_k\hat{q}_k + \frac{1}{2}g_k E_k h_k s \in \mathbb{R}^{2r}, \tag{8.21}$$

where $\varsigma > 0$, and $\zeta > 0$ and $\tau > 0$ are some properly selected scalars, $\hat{q}_k(0) = 0$, $\kappa_k > 0$ and matrix $E_k \in \mathbb{R}^{2r \times 2r}$.

Proof For the adaptive dynamic SMC law (8.20), we consider the Lyapunov function

$$L = \frac{1}{2}s^2 + \frac{1}{2}\sum_{k=1}^{2}\tilde{q}_k^{\mathrm{T}}E_k^{-1}\tilde{q}_k.$$

Then, from (8.14), (8.19) and (8.20), it follows that

$$
\begin{aligned}
\dot{L} &= s\dot{s} - \sum_{k=1}^{2}\tilde{q}_k^{\mathrm{T}}E_k^{-1}\dot{\tilde{q}}_k = s(g\dot{e} + h\dot{v}) - \sum_{k=1}^{2}\tilde{q}_k^{\mathrm{T}}E_k^{-1}(-\kappa_k E_k\hat{q}_k + \frac{1}{2}g_k E_k h_k s) \\
&= sg\Big[(A - v_0\hat{B})e - \hat{B}ev + \bar{Q} + \hat{B}x_{\mathrm{ref}}(v + v_0) + W\Big] - sg\Big(Ae + \hat{B}xv \\
&\quad + v_0\hat{B}x + W\Big) - \frac{1}{2}\sum_{k=1}^{2}sg_k\hat{q}_k^{\mathrm{T}}h_k - \varsigma s^2 - \zeta s|s|^{\tau}\mathrm{sign}(s) \\
&\quad - \sum_{k=1}^{2}\tilde{q}_k^{\mathrm{T}}(-\kappa_k\hat{q}_k + \frac{1}{2}g_k E_k h_k s) \\
&= s\begin{bmatrix} g_1 & g_2 \end{bmatrix}\begin{bmatrix} q_1 & q_2 \end{bmatrix}^{\mathrm{T}} - \frac{1}{2}\sum_{k=1}^{2}sg_k\hat{q}_k^{\mathrm{T}}h_k - \varsigma s^2 - \zeta|s|^{1+\tau} + \sum_{k=1}^{2}\kappa_k\tilde{q}_k^{\mathrm{T}}\hat{q}_k \\
&\quad - \frac{1}{2}\sum_{k=1}^{2}g_k\tilde{q}_k^{\mathrm{T}}h_k s \\
&= s\cdot\sum_{k=1}^{2}g_k(\frac{1}{2}q_k^{\mathrm{T}}h_k + \delta_k) - \frac{1}{2}\sum_{k=1}^{2}g_k\hat{q}_k^{\mathrm{T}}h_k s - \varsigma s^2 - \zeta|s|^{1+\tau} + \sum_{k=1}^{2}\kappa_k\tilde{q}_k^{\mathrm{T}}\hat{q}_k \\
&\quad - \frac{1}{2}\sum_{k=1}^{2}g_k\tilde{q}_k^{\mathrm{T}}h_k s \\
&= s(g_1\delta_1 + g_2\delta_2) - \varsigma s^2 - \zeta|s|^{1+\tau} + \sum_{k=1}^{2}\kappa_k\tilde{q}_k^{\mathrm{T}}\hat{q}_k.
\end{aligned}
$$

Meanwhile, for the vector $\hat{q}_k = q_k - \tilde{q}_k$, considering $2\tilde{q}_k^{\mathrm{T}}\hat{q}_k \leq -\|\tilde{q}_k\|^2 + \|q_k\|^2$ [28] and the fact that $s\cdot(g_1\delta_1 + g_2\delta_2) \leq \theta\cdot s^2 + \frac{1}{\theta}\cdot(g_1\delta_1 + g_2\delta_2)^2$ with $\theta > 0$, we can further derive $\dot{L}$ as follows:

$$
\begin{aligned}
\dot{L} &\leq \theta s^2 + \frac{1}{\theta}(g_1\delta_1 + g_2\delta_2)^2 - \varsigma s^2 - \zeta|s|^{1+\tau} - \frac{1}{2}\sum_{k=1}^{2}\kappa_k\|\tilde{q}_k\|^2 + \frac{1}{2}\sum_{k=1}^{2}\kappa_k\|q_k\|^2 \\
&= -(\varsigma - \theta)s^2 - \zeta|s|^{1+\tau} - \frac{1}{2}\sum_{k=1}^{2}\kappa_k\|\tilde{q}_k\|^2 + \frac{1}{2}\sum_{k=1}^{2}\kappa_k\|q_k\|^2,
\end{aligned}
$$

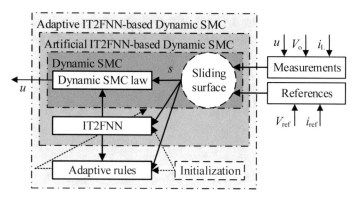

Fig. 8.2 Structure of the designed adaptive IT2FNN-based DSMC

from which one can always find solutions to $(\varsigma - \theta) > \frac{1}{2}$ and $\frac{\kappa_k}{2} > \frac{1}{\lambda_{\max}(E_k^{-1})}$, that is to say $\dot{L} \leq -\varphi L + \phi$ with $\varphi > 0$ and $\phi = -\zeta|s|^{1+\tau} + \frac{1}{2}(\kappa_1\|\boldsymbol{q}_1\|^2 + \kappa_2\|\boldsymbol{q}_2\|^2)$ holds for any $s \neq 0$. Since ϕ is bounded, the bounded stability of L can be thus ensured. This completes the proof. ∎

Remark 8.3 Since the condition of $s = 0$ (in sliding stage) and $\hat{\boldsymbol{q}}_k(0) = 0$ implies the dynamic SMC law is same with the one in Theorem 8.2, the analysis of the existence of the desired sliding motion is similar to Theorem 8.3, and is omitted here.

Summarizing from the designed control laws based on the measurements $\{i_L, V_o, u\}$ and reference inputs $\{i_{\text{ref}}, V_{\text{ref}}\}$, the structure of the designed adaptive IT2FNN-based dynamic SMC is illustrated in Fig. 8.2, in which the output u is the desired duty cycle of the converter in Fig. 8.1. Further, in order to design an adaptive IT2FNN for a wider utilization on the elimination of various uncertainties, such as different unknown perturbations caused by input voltage V_i, input inductor L, the load resistor R, output capacitor C, and noises, we design some adaptive rules for the updating of the parameters m_{ij}, v_{ij} and α_{ij} of the designed type-2 FNN.

8.3.3 Design of the Parameter Updating Rules of the Ellipsoidal-Type IT2FNN

In terms of the antecedent parameters m_{ij}, v_{ij} and α_{ij} used in each fuzzy rule of the IT2FNN, we apply the GDM to update m_{ij}, v_{ij} and α_{ij} for the utilization in the flexible estimation of the uncertainties. Let m_{ij}, v_{ij} and α_{ij} be expressed by ξ. The following gradient descent algorithm is designed for the parameter updating rules:

$$\xi(k + 1) = \xi(k) - \mu \cdot \frac{\partial \epsilon}{\partial \xi}, \tag{8.22}$$

where $\epsilon = \frac{1}{2}e^2$ (e here denotes e_i or e_V), $\frac{\partial \epsilon}{\partial \xi}$ denotes a partial derivative of the function ϵ of several variables ξ (m_{ij}, v_{ij} and α_{ij}). Meanwhile, we use the following updating law, that is,

$$\frac{\partial \epsilon}{\partial \xi} = \frac{1}{2} \cdot \left(\frac{\partial m_L}{\partial \underline{h}_i} \cdot \frac{\partial \underline{h}_i}{\partial \underline{g}_{ij}} \cdot \frac{\partial \underline{g}_{ij}}{\partial \xi} + \frac{\partial m_U}{\partial \bar{h}_i} \cdot \frac{\partial \bar{h}_i}{\partial \bar{g}_{ij}} \cdot \frac{\partial \bar{g}_{ij}}{\partial \xi} \right) \cdot e$$

to update $\frac{\partial \epsilon}{\partial \xi}$, where

$$\frac{\partial m_L}{\partial \underline{h}_i} = \frac{\hat{q}_{Li} - m_L(\hat{q}_{Li}, \underline{h}_i)}{\sum_{l=1}^r \underline{h}_l}, \quad \frac{\partial m_U}{\partial \bar{h}_i} = \frac{\hat{q}_{Ui} - m_U(\hat{q}_{Ui}, \bar{h}_i)}{\sum_{l=1}^r \bar{h}_l},$$

$$\frac{\partial \underline{h}_i}{\partial \underline{g}_{ij}} = \prod_{l=1,l\neq j}^3 \underline{g}_{il}, \quad \frac{\partial \bar{h}_i}{\partial \bar{g}_{ij}} = \prod_{l=1,l\neq j}^3 \bar{g}_{il},$$

and $\frac{\partial \underline{g}_{ij}}{\partial \xi}$ and $\frac{\partial \bar{g}_{ij}}{\partial \xi}$ are provided in Table 8.1.

Remark 8.4 Actually, the designed parameter updating rules of the ellipsoidal-type IT2FNN can be utilized to a wider ellipsoidal-type IT2FNN with more inputs, although there are only three inputs to the IT2FNN. The number of the fuzzy rules is related to the number of the fuzzy sets chosen for each input z_j. More fuzzy sets imply more complex computation, but more precise tracking control of the voltage. Therefor, a trade-off between the selection of fuzzy sets and the accurate voltage regulation is recommended. In the simulation subsequently, we use two types of type-2 fuzzy sets for each input to achieve the voltage regulation.

Table 8.1 $\frac{\partial \underline{g}_{ij}}{\partial \xi}$ and $\frac{\partial \bar{g}_{ij}}{\partial \xi}$ of the ellipsoidal-type IT2 membership functions

	$m_{ij} - v_{ij} < z_i \leq m_{ij}$	$m_{ij} < z_i \leq m_{ij} + v_{ij}$
$\frac{\partial \underline{g}_{ij}}{\partial \alpha_{ij}}$	$-\left(\frac{m_{ij}-x_i}{v_{ij}} \right)^{\alpha_{ij}} \ln\left(\frac{m_{ij}-x_i}{v_{ij}} \right)$	$\left(\frac{x_i-m_{ij}}{v_{ij}} \right)^{\alpha_{ij}} \ln\left(\frac{x_i-m_{ij}}{v_{ij}} \right)$
$\frac{\partial \bar{g}_{ij}}{\partial \alpha_{ij}}$	$\frac{1}{\alpha_{ij}^2} \left(\frac{m_{ij}-x_i}{v_{ij}} \right)^{\frac{1}{\alpha_{ij}}} \ln\left(\frac{m_{ij}-x_i}{v_{ij}} \right)$	$-\frac{1}{\alpha_{ij}^2} \left(\frac{x_i-m_{ij}}{v_{ij}} \right)^{\frac{1}{\alpha_{ij}}} \ln\left(\frac{x_i-m_{ij}}{v_{ij}} \right)$
$\frac{\partial \underline{g}_{ij}}{\partial m_{ij}}$	$-\frac{\alpha_{ij}}{v_{ij}} \left(\frac{m_{ij}-x_i}{v_{ij}} \right)^{\alpha_{ij}-1}$	$\frac{\alpha_{ij}}{v_{ij}} \left(\frac{x_i-m_{ij}}{v_{ij}} \right)^{\alpha_{ij}-1}$
$\frac{\partial \bar{g}_{ij}}{\partial m_{ij}}$	$-\frac{1}{\alpha_{ij}v_{ij}} \left(\frac{m_{ij}-x_i}{v_{ij}} \right)^{\frac{1-\alpha_{ij}}{\alpha_{ij}}}$	$\frac{1}{\alpha_{ij}v_{ij}} \left(\frac{x_i-m_{ij}}{v_{ij}} \right)^{\frac{1-\alpha_{ij}}{\alpha_{ij}}}$
$\frac{\partial \underline{g}_{ij}}{\partial v_{ij}}$	$\frac{\alpha_{ij}}{v_{ij}} \left(\frac{m_{ij}-x_i}{v_{ij}} \right)^{\alpha_{ij}}$	$-\frac{\alpha_{ij}}{v_{ij}} \left(\frac{x_i-m_{ij}}{v_{ij}} \right)^{\alpha_{ij}}$
$\frac{\partial \bar{g}_{ij}}{\partial v_{ij}}$	$\frac{1}{\alpha_{ij}v_{ij}} \left(\frac{m_{ij}-x_i}{v_{ij}} \right)^{\frac{1}{\alpha_{ij}}}$	$-\frac{1}{\alpha_{ij}v_{ij}} \left(\frac{x_i-m_{ij}}{v_{ij}} \right)^{\frac{1}{\alpha_{ij}}}$

8.4 Simulation Results

In this study, we numerically simulate the voltage regulation of the considered converter by using the average model (8.1) and (8.2). The uncertainties in the model caused by the perturbations of the input inductor and output capacitor the, as well as the variations of the input voltage and load resistance, are fully considered in simulations. This section will present simulation results by three cases, i.e., (i) control against the input voltage variation, (ii) control against the load resistance variation, and (iii) control against both the variations of the input voltage and load resistance. For a comparative illustration of the proposed control method, the designed dynamic SMC, artificial IT2FNN-based dynamic SMC, and adaptive IT2FNN-based dynamic SMC are applied for each of the three cases.

8.4.1 Set Up

According to the configuration of the DC/DC boost converter, we take the nominal values of the model parameters from Table 8.2.

In simulations, we set the sampling time at $T = 50$ μs. Besides, the prototype of the converter is considered with a nominal output power $P_o = 480$ W, while the high-voltage bus is set at $V_{ref} = 120$ V for some inverter utilization. Accordingly, the reference input of the inductor current is designed as $i_{ref} = \kappa_p e_1 + \kappa_i \int e_1 + \kappa_d \dot{e}_1$ with $\kappa_p = 0.12$, $\kappa_i = 3310$ and $\kappa_d = 0.005$ [27]. Based on these setups, we use the designed control laws to illustrate the effectiveness and robustness of the proposed voltage tracking control method in different cases, subsequently.

8.4.2 Case 1: Control Against the Input Voltage Variation

Since the nominal value of the input voltage is $V_{in} = 60$ V, we might consider that V_i is varied between $48 \le V_i \le 60$ and then between $60 \le V_i \le 72$ with the load resistance $R = 30\,\Omega$ ($P_o = 480$ W) being fixed. A varied V_i is depicted in Fig. 8.3. In the following, we use respectively the dynamic SMC law (8.17), the IT2FNN-based

Table 8.2 Nominal values of the parameters of the DC/DC boost converter

Parameters	Values	Units	Description
C_n	860	μF	Output capacitor
f_s	20	kHz	Switching frequency
L_n	860	μH	Input inductor
R_n	30	Ω	Load resistance
V_{in}	60	V	Input voltage

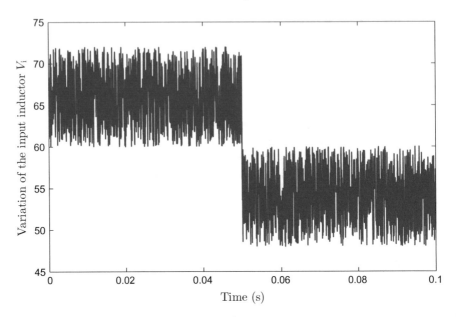

Fig. 8.3 Variation of the input inductor V_i

dynamic SMC law (8.20) with artificial parameters m_{ij}, v_{ij} and α_{ij}, and IT2FNN-based dynamic SMC law (8.20) with adaptive parameters m_{ij}, v_{ij} and α_{ij} to regulate the output voltage of the DC/DC boost converter.

In this case, specifically, according to Theorems 8.1 and 8.2, the parameters of the dynamic SMC law (8.17) are set to

$$g = \begin{bmatrix} 0.0037 & 0.0052 \end{bmatrix}, \; h = 850.0, \; \varsigma = 0.0470, \; \zeta = 0.10, \; \tau = 0.10. \quad (8.23)$$

Then, according to Theorem 8.3, the parameters of the IT2FNN-based dynamic SMC law (8.20) are designed as (8.23) with updated $\varsigma = 0.0470$. Besides, for each input z_j of the FNN, we use two types of IT2 fuzzy sets, i.e., PS (Positive Small) and PB (Positive Big). The corresponding membership functions $g_{ij}(z_j)$ with FOUs are shown in Fig. 8.4, where the artificial parameters m_{ij}, v_{ij} and α_{ij} for the membership functions $\underline{g}_{ij}(z_j)$ and $\bar{g}_{ij}(z_j)$ in (8.10)–(8.11) are chosen as:

$$
\begin{aligned}
m_{\{1,2,3,4\}1} &= 3.9583, & m_{\{5,6,7,8\}1} &= 6.4583, \\
m_{\{1,2,5,6\}2} &= 78, & m_{\{3,4,7,8\}2} &= 102, \\
m_{\{1,3,5,7\}3} &= 0.3, & m_{\{2,4,6,8\}3} &= 0.7, \\
v_{i1} &= 1.8750, & v_{i2} &= 18, \\
v_{i3} &= 0.3, & \alpha_{i1} &= 0.5, \\
\alpha_{i2} &= 0.4, & \alpha_{i3} &= 0.6, \; i = 1, 2, \ldots, 8. \quad (8.24)
\end{aligned}
$$

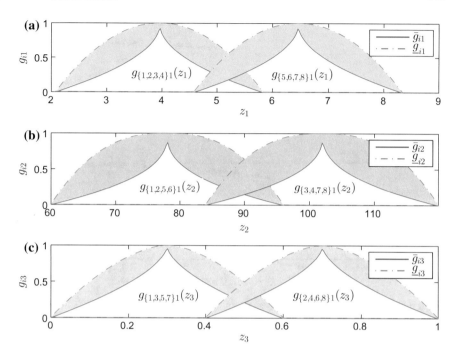

Fig. 8.4 MFs of IT2FNN for inputs z_1, z_2 and z_3

Further, we design the adaptive IT2FNN to obtain the IT2FNN-based dynamic SMC law (8.20). The controller parameters are tuned as (8.23) with updated $\varsigma = 0.0528$. The parameters m_{ij}, v_{ij} and α_{ij} for the lower and upper membership functions are updated according to the adaptive rules in Sect. 8.3.3. The initial conditions of the parameters are set as the ones in (8.24). We omit the computational process of the adaptive parameters, which are calculated by the gradient descent algorithm (8.22).

For simplicity, we denote the three control approaches, i.e., the dynamic SMC law (8.17), artificial IT2FNN-based dynamic SMC law (8.20) and adaptive IT2FNN-based dynamic SMC law (8.20), respectively, the Control A, the Control B and the Control C. In this case, these resulting duty cycle inputs u are depicted in Fig. 8.5a. For the Controls B and C, the parameters of the adaptive laws (8.21) are set to $\kappa_1 = 0.1$, $\kappa_2 = 0.2$ and $E_1 = E_2 = I_{16}$. Figure 8.5b shows the trajectories of the corresponding sliding surface s (8.14). According to these three control laws under the considered case, the regulations of the capacitor voltage V_o are shown in Fig. 8.6. Figure 8.7 illustrates the tracking errors of the capacitor voltage. Besides, the inductor current and its tracking error are respectively shown in Figs. 8.8 and 8.9 for the three control approaches. From the simulation results, we know that it illustrates the robustness of the three control algorithms to the variation of the input voltage, although no information of the uncertainties is needed to the IT2FNN-based dynamic SMC law (8.20).

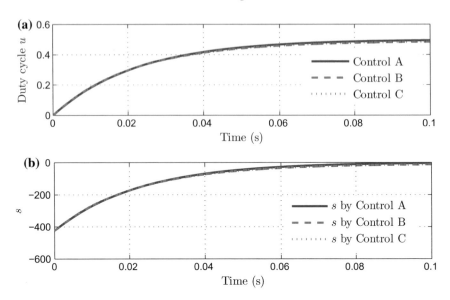

Fig. 8.5 Duty cycle and sliding surface under Case 1 by the three controls

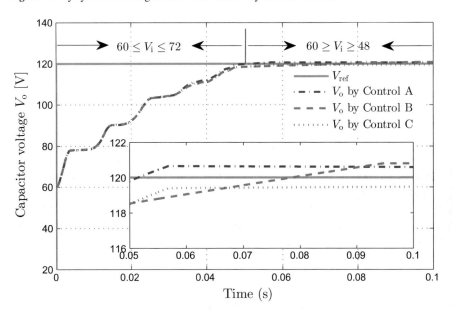

Fig. 8.6 Capacitor voltage under Case 1 by the three controls

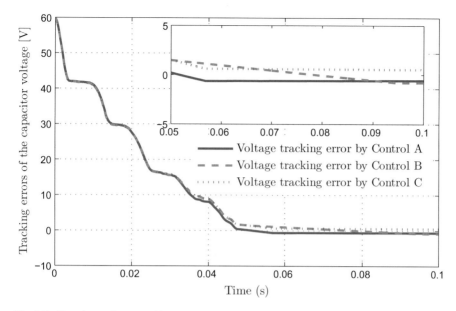

Fig. 8.7 Capacitor voltage tracking error under Case 1 by the three controls

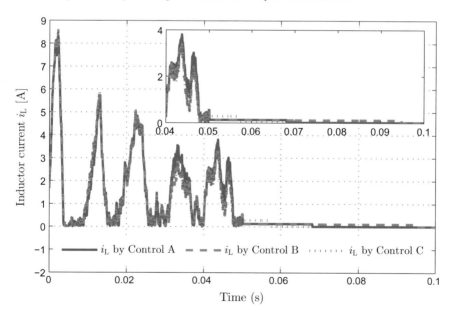

Fig. 8.8 Inductor current under Case 1 by the three controls

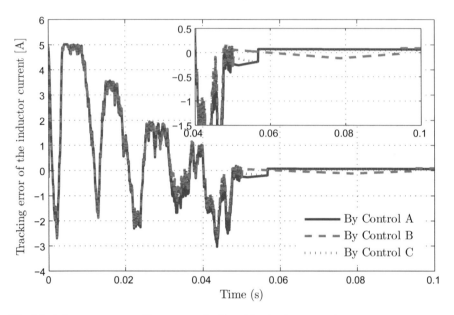

Fig. 8.9 Inductor current tracking error under Case 1 by the three controls

8.4.3   Case 2: Control Against the Load Resistance Variation

In this case, the load resistance R is supposed to be varied between $5 \leq R \leq 123 \, \Omega$ ($200 \leq P_o \leq 800 \, \text{W}$), and the input voltage is set at $V_{\text{in}} = 60 \, \text{V}$. A piecewise variation of R is depicted in Fig. 8.10. Also, we use the three control approaches to verify the robustness to the load resistance variation. The parameters of the three controllers are given as follows:

$$g = \begin{bmatrix} 0.0037 & 0.0052 \end{bmatrix}, \; h = 800.0, \; \varsigma = 0.335, \; \zeta = 0.15, \; \tau = 0.10. \quad (8.25)$$

For Control C, we update $\varsigma = 0.345$ and use the same initial conditions as presented in (8.24) for the IT2FNN. The resulting duty cycle inputs u are shown in Fig. 8.11a. Figure 8.11b depicts the corresponding curves of the three sliding surfaces.

By these three control laws, the regulations of the capacitor voltage V_o, in this case, are illustrated in Fig. 8.12, and the tracking errors of the capacitor voltage are shown in Fig. 8.13. Figures 8.14 and 8.15 show the inductor current and its tracking error, respectively. Obviously, the robust control of the DC/DC boost converter with the load resistance variation performs well by the designed control approaches.

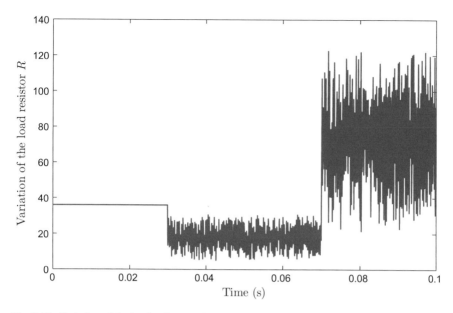

Fig. 8.10 Variation of the load resistance R

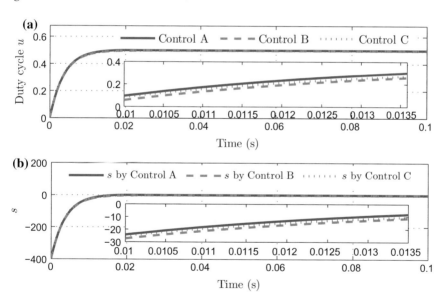

Fig. 8.11 Duty cycle and sliding surface under Case 2 by the three controls

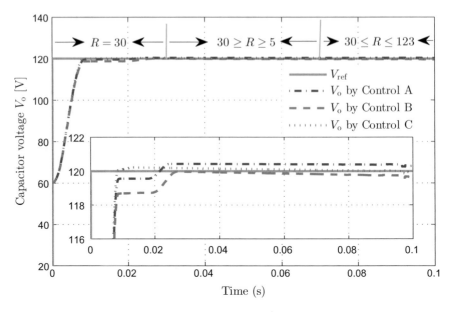

Fig. 8.12 Capacitor voltage under Case 2 by the three controls

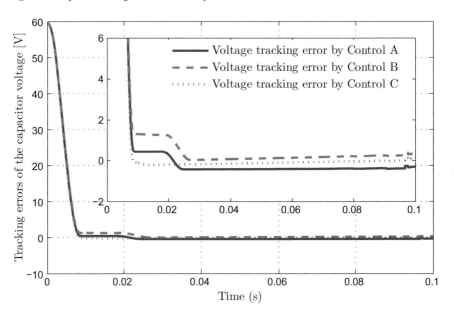

Fig. 8.13 Capacitor voltage tracking error under Case 2 by the three controls

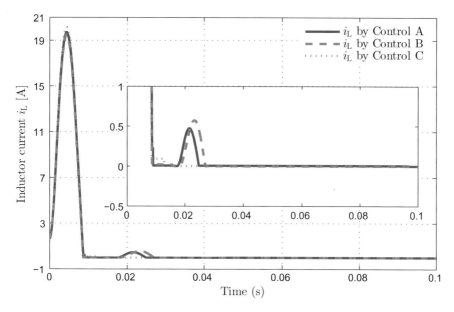

Fig. 8.14 Inductor current under Case 2 by the three controls

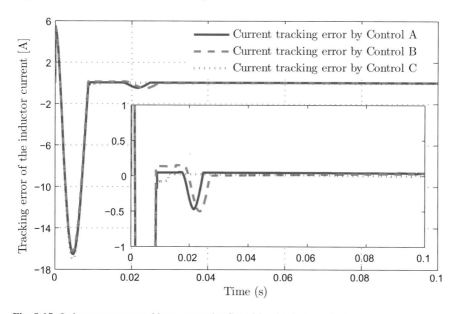

Fig. 8.15 Inductor current tracking error under Case 2 by the three controls

8.4.4 Case 3: Control Against the Variations of the Load Resistance and Input Voltage

Generally, we consider the perturbations from both the load resistance and input voltage in this case. The variations of V_i and R are used exactly as the ones in Cases 1 and 2. The parameters of the dynamic SMC law are given as:

$$g = \begin{bmatrix} 0.0037 & 0.0052 \end{bmatrix}, \ h = 800.0, \ \varsigma = 0.0468, \ \zeta = 0.10, \ \tau = 0.10. \quad (8.26)$$

Moreover, we update $\varsigma = 0.0558$ for Controls B and C for good tracking performance, and the initial conditions for the adaptive IT2FNN are the same as the ones used in the first two cases.

Accordingly, the resulting duty cycle inputs u are shown in Fig. 8.16a. Figure 8.16b depicts the curves of the sliding surfaces. In this case, the regulations of the capacitor voltage V_o are shown in Fig. 8.17, while Fig. 8.18 depicts the tracking errors of the capacitor voltage. Figures 8.19 and 8.20 illustrate the inductor current and its tracking error, respectively. Evidently, the three control scheme is effective to the voltage regulation of the DC/DC boost converter. Moreover, from the simulation results of the three cases, it can be concluded that the designed adaptive IT2FNN-based dynamic SMC scheme is more suitable without chattering in the control efforts for the voltage regulation of the converter, despite with unknown information of the variation of the system parameters.

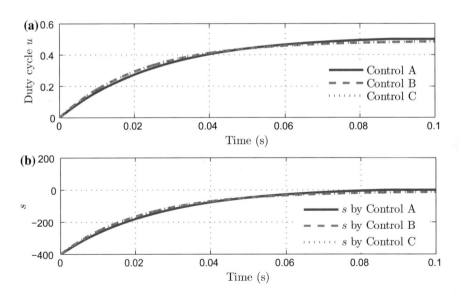

Fig. 8.16 Duty cycle and sliding surface under Case 3 by the three controls

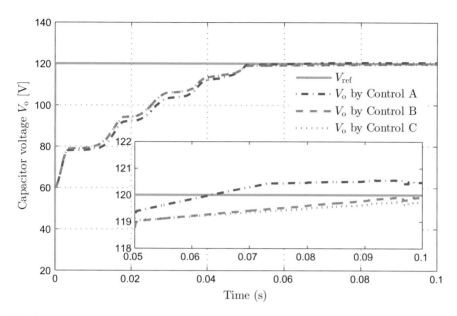

Fig. 8.17   Capacitor voltage under Case 3 by the three controls

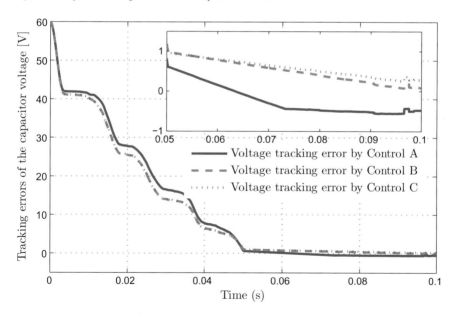

Fig. 8.18   Capacitor voltage tracking error under Case 3 by the three controls

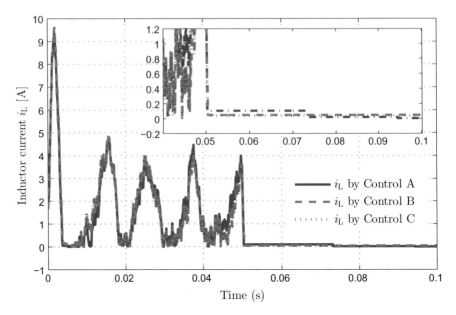

Fig. 8.19 Inductor current under Case 3 by the three controls

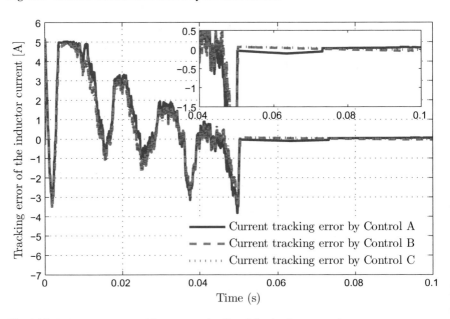

Fig. 8.20 Inductor current tracking error under Case 3 by the three controls

8.5 Conclusion

The problem of robust voltage regulation of the DC/DC boost converter has been solved via the IT2FNN-based SMC in this paper. A modified linear sliding surface has been designed, which is depending on the desired voltage and the capacitor voltage. A new dynamic SMC against the complex uncertainties has been designed. Moreover, artificial and adaptive IT2FNN-based dynamic SMC laws have been synthesized. The updating rules of the parameters of simplified ellipsoidal-type membership functions have been designed based on the GDM. In the end, simulation results have verified the effectiveness of the presented adaptive IT2FNN-based SMC method. To accurately estimate the uncertainties existing the dynamics of the converter, some disturbance-observer-based control techniques can be considered for the voltage regulation of DC/DC boost converters.

References

1. Chen, C.L.P., Wen, G.X., Liu, Y.J., Wang, F.Y.: Adaptive consensus control for a class of nonlinear multiagent time-delay systems using neural networks. IEEE Trans. Neural Netw. Learn. Syst. **25**(6), 1217–1226 (2014)
2. Dijk, E.V., Spruijt, J., O'sullivan, D.M., Klaassens, J.B.: PWM-switch modeling of DC-DC converters. IEEE Trans. Power Electron. **10**(6), 659–665 (1995)
3. Doncker, R.W.D., Divan, D.M., Kheraluwala, M.H.: A three-phase soft-switched high-power-density DC/DC converter for high-power applications. IEEE Trans. Ind. Appl. **27**(1), 63–73 (1991)
4. Gao, Q., Zeng, X.J., Feng, G., Wang, Y., Qiu, J.: T-S-fuzzy-model-based approximation and controller design for general nonlinear systems. IEEE Trans. Syst. Man Cybern. Part B: Cybern. **42**(4), 1143–1154 (2012)
5. Gao, W., Hung, J.C.: Variable structure control of nonlinear systems: a new approach. IEEE Trans. Ind. Electron. **40**(1), 45–55 (1993)
6. Juang, C.F., Huang, R.B., Cheng, W.Y.: An interval type-2 fuzzy-neural network with support-vector regression for noisy regression problems. IEEE Trans. Fuzzy Syst. **18**(4), 686–699 (2010)
7. Kazimierczuk, M.K.: Pulse-width Modulated DC-DC Power Converters. Wiley (2015)
8. Khanesar, M.A., Kayacan, E., Teshnehlab, M., Kaynak, O.: Extended Kalman filter based learning algorithm for type-2 fuzzy logic systems and its experimental evaluation. IEEE Trans. Ind. Electron. **59**(11), 4443–4455 (2012)
9. Lam, H., Xiao, B., Yu, Y., Yin, X., Han, H., Tsai, S.H., Chen, C.S.: Membership-function-dependent stability analysis and control synthesis of guaranteed cost fuzzy-model-based control systems. Int. J. Fuzzy Syst. **18**(4), 537–549 (2016)
10. Lam, H.K., Li, H., Deters, C., Secco, E.L., Wurdemann, H.A., Althoefer, K.: Control design for interval type-2 fuzzy systems under imperfect premise matching. IEEE Trans. Ind. Electron. **61**(2), 956–968 (2014)
11. Lee, J.Y., Jeong, Y.S., Han, B.M.: An isolated DC/DC converter using high-frequency unregulated LLC resonant converter for fuel cell applications. IEEE Trans. Ind. Electron. **58**(7), 2926–2934 (2011)
12. Lin, Y.Y., Chang, J.Y., Lin, C.T.: A TSK-type-based self-evolving compensatory interval type-2 fuzzy neural network (TSCIT2FNN) and its applications. IEEE Trans. Ind. Electron. **61**(1), 447–459 (2014)

13. Lin, Y.Y., Liao, S.H., Chang, J.Y., Lin, C.T.: Simplified interval type-2 fuzzy neural networks. IEEE Trans. Neural Netw. Learn. Syst. **25**(5), 959–969 (2014)
14. Liu, J., Gao, Y., Luo, W., Wu, L.: Takagi-Sugeno fuzzy-model-based control of three-phase AC/DC voltage source converters using adaptive sliding mode technique. IET Control Theory Appl. **11**(8), 1255–1263 (2016)
15. Liu, J., Gao, Y., Su, X., Wack, M., Wu, L.: Disturbance-observer-based control for air management of PEM fuel cell systems via sliding mode technique. IEEE Trans. Control Syst. Technol. **27**(3), 1129–1138 (2019)
16. Liu, J., Yin, Y., Luo, W., Vazquez, S., Franquelo, L.G.: Sliding mode control of a three-phase AC/DC voltage source converter under unknown load conditions: industry applications. IEEE Trans. Syst. Man Cybern.: Syst. **48**(10), 1771–1780 (2018)
17. Mendel, J.M.: Uncertain rule-based fuzzy systems. In: Introduction and New Directions, p. 684. Springer (2017)
18. Mendez-Diaz, F., Pico, B., Vidal-Idiarte, E., Calvente, J., Giral, R.: HM/PWM seamless control of a bidirectional buck-boost converter for a photovoltaic application. IEEE Trans. Power Electron. **34**(3), 2887–2899 (2019)
19. Morroni, J., Corradini, L., Zane, R., Maksimovic, D.: Adaptive tuning of switched-mode power supplies operating in discontinuous and continuous conduction modes. IEEE Trans. Power Electron. **24**(11), 2603–2611 (2009)
20. Oucheriah, S., Guo, L.: PWM-based adaptive sliding-mode control for boost DC-DC converters. IEEE Trans. Ind. Electron. **60**(8), 3291–3294 (2013)
21. Shen, L., Lu, D.D.C., Li, C.: Adaptive sliding mode control method for DC-DC converters. IET Power Electron. **8**(9), 1723–1732 (2015)
22. Sira-Ramirez, H.: Sliding-mode control on slow manifolds of DC-to-DC power converters. Int. J. Control **47**(5), 1323–1340 (1988)
23. Thounthong, P., Raël, S., Davat, B.: Control strategy of fuel cell and supercapacitors association for a distributed generation system. IEEE Trans. Ind. Electron. **54**(6), 3225–3233 (2007)
24. Utkin, V., Gulder, J., Shi, J.: Sliding mode control in electro-mechanical systems. In: Automation and Control Engineering Series, vol. 34. Taylor & Francis Group (2009)
25. Wai, R.J., Lin, Y.F., Liu, Y.K.: Design of adaptive fuzzy-neural-network control for a single-stage boost inverter. IEEE Trans. Power Electron. **30**(12), 7282–7298 (2015)
26. Wai, R.J., Shih, L.C.: Design of voltage tracking control for DC-DC boost converter via total sliding-mode technique. IEEE Trans. Ind. Electron. **58**(6), 2502–2511 (2011)
27. Wai, R.J., Shih, L.C.: Adaptive fuzzy-neural-network design for voltage tracking control of a DC-DC boost converter. IEEE Trans. Power Electron. **27**(4), 2104–2115 (2012)
28. Wang, D., Huang, J.: Neural network-based adaptive dynamic surface control for a class of uncertain nonlinear systems in strict-feedback form. IEEE Trans. Neural Netw. **16**(1), 195–202 (2005)
29. Wang, J.: A new type of fuzzy membership function designed for interval type-2 fuzzy neural network. Acta Autom. Sin. **43**(8), 1425–1433 (2017)
30. Wu, L., Gao, Y., Liu, J., Li, H.: Event-triggered sliding mode control of stochastic systems via output feedback. Automatica **48**, 79–92 (2017)
31. Yang, W.H., Huang, C.J., Huang, H.H., Lin, W.T., Chen, K.H., Lin, Y.H., Lin, S.R., Tsai, T.Y.: A constant-on-time control DC-DC buck converter with the pseudowave tracking technique for regulation accuracy and load transient enhancement. IEEE Trans. Power Electron. **33**(7), 6187–6198 (2018)
32. Yin, Y., Liu, J., Sanchez, J.A., Wu, L., Vazquez, S., Leon, J.I., Franquelo, L.G.: Observer-based adaptive sliding mode control of NPC converters: an RBF neural network approach. IEEE Trans. Power Electron. **34**(4), 3831–3841 (2019)

Chapter 9
Sliding Mode Control of Three-Phase Power Converters

This chapter presents an ESO-based SOSM control for three-phase two-level grid-connected power converters. The presented control technique forces the input currents to track the desired values, which can indirectly regulate the output voltage while achieving a user-defined power factor. The presented approach has two control loops. A current control loop based on an SOSM and a DC-link voltage regulation loop which consists of an ESO plus SOSM. The load connected to the DC-link capacitor is considered as an external disturbance. An ESO is used to asymptotically reject this external disturbance. Therefore, its design is considered in the control law derivation to achieve a high performance. Theoretical analysis is given to show the closed-loop behavior of the proposed controller and experimental results are presented to validate the control algorithm under a real power converter prototype.

9.1 Introduction

Over the last decades, power converters have experienced a tremendous development in industrial applications, mainly due to their increased reliability, their advantageous properties such as high efficiency and power capacity, the reduced cost of power-electronic devices, and the increased application demands in industry [17]. Among the power converters, the three-phase PWM power converters are most commonly used, whose applications can be found in motor drives, energy storage systems, integration of renewable energy sources, etc. [33, 37].

Three-phase PWM converters play a key role in industrial applications like integration of renewable energy sources, energy storage systems, motor drives, etc [3, 26]. Particularly, active front ends (AFE) are grid-connected converters that offer features as bidirectional power flow, near-sinusoidal currents, and power factor and DC-link capacitor voltage regulation capability [35]. For this reason, the control

© Springer Nature Switzerland AG 2020
J. Liu et al., *Sliding Mode Control Methodology in the Applications of Industrial Power Systems*, Studies in Systems, Decision and Control 249,
https://doi.org/10.1007/978-3-030-30655-7_9

objectives for this application are: (1) maintain the DC-link voltage regulated to a certain reference, and (2) supply a desired reactive power and draw grid currents with the lowest possible harmonic distortion.

To make the AFEs achieve good performance, there are many approaches to develop the control law for this system. Early solutions were linear regulators like the one proposed in [25]. It changes the modulation index slowly to regulate the DC-link capacitor voltage. Therefore, its main disadvantage is that it presents a slow dynamical response. A faster response can be obtained using a deadbeat current control [38]. However, it is well known that it is highly sensitive to the parameter uncertainties. In general, these approaches define an operating point and then work with a small-signal linearized model around it. This is a drawback because they can not guarantee stability against large signal disturbances for the large range of operating conditions of the three-phase PWM converter due to the fact that AFE are nonlinear systems. Furthermore, the controller implementation requires the system parameters which depend on the operating points, otherwise it may result in steady state errors in the state variables. On the other hand, in light of the strong nonlinearity of AFE, nonlinear control algorithms may be suitable for controlling the power converters, which are able to accommodate a wide range of operating conditions. The main reason is that there's no need to have a linear model of the power converter for nonlinear controller design.

Several nonlinear control approaches have been proposed for grid-tied power converters, such as input-output linearization [16], nonlinear adaptive control [20], passivity based control [5, 10], feedback linearization [15], differential flatness based control [11, 24, 31], model based predictive control [33, 35] and SMC [30]. Among these techniques, SMC is suitable for dealing with the nonlinear behavior of the considered system due to its characteristics of insensitivity to external disturbances, system reduction, high accuracy and finite time convergence [7, 19].

SMC has been developed as a new control design method for a wide spectrum of systems including nonlinear, time-varying and fault tolerant systems [12]. SMC can manage the nonlinear behavior of the three-phase grid-connected power converter. Besides, it is characterized to be a robust and effective control strategy. However, up to authors knowledge, proposed SMC strategies have only considered the control of input current in sliding mode [21, 27]. In general, these works use a proportional integral (PI) controller for the DC-link capacitor voltage regulation. This approach achieves robustness of input current control, but can't guarantee robustness of output voltage control since they are derived using approximations and linearizations. To solve this issue, extended state observer (ESO) based control strategies are proposed for voltage control design to reject disturbances and uncertainties [22, 34]. ESO is an efficient technique for disturbance estimation, which regards the lumped disturbances (such as parametric uncertainty, unmodeled dynamics and load variation) as a new state. However, only simulation results are provided in both works, and the stability analysis of the closed-loop system with modeling uncertainties is not performed.

In this chapter, a model-based second-order SMC for three-phase grid-connected power converters is proposed. The controller design is based on the system model and has a cascaded structure which consists of two control loops. The outer loop regulates the DC-link capacitor voltage and the power factor providing the current references for the inner control loop. The current control loop tracks the actual currents to their desired values. To design the proposed controller, the load connected to the DC-link capacitor is considered as a disturbance, which directly affects the performance of the whole system. Thus, a composite control law consisting of SOSM based on STA and disturbance compensation via ESO is developed for the voltage regulation loop. The current control loop is also designed using the STA. The STA is one of the most popular SOSM algorithms and a unique absolutely continuous sliding mode algorithm, ensuring all the main properties of first order SMC for systems with Lipschitz continuous matched uncertainties/disturbances with bounded gradients [8]. The modeling uncertainties are incorporated into the design of ESO and SOSM algorithms. The sliding surface is designed ensuring the finite time asymptotic convergence of the sliding variables to its desired values in the presence of parametric uncertainties. In addition, theoretical study of the stability issue is provided using Lyapunov method.

To show the benefits of the proposed controller it can be compared with some conventional control strategies like PI synchronous reference frame (PI-SRF) or proportional plus resonant (PR) controllers. The PI-SRF is considered as the standard control method in industry for this system and has been finally selected as the baseline controller. It should be noticed that the PR controller is also a linear control. Besides, when PR is adopted, a simple PI regulator is usually employed for the outer control loop. Therefore, it is expected that proposed ESO-SOSM provides the same improvements as for the PI-SRF. That is, better transient response and improved system performance under a load step. The performance of the proposed control is compared with a well-tuned conventional PI-SRF controller. The results show that the performance of the designed controller presents a faster dynamic behavior while maintaining a lower total harmonic distortion (THD) value in steady state and improves system performance under a load step. The validity of the proposed control algorithm has been verified by experimental results for a 3.0 kVA insulated gate bipolar transistor (IGBT) PWM power converter using a TMS320F28335 digital signal processor (DSP).

9.2 Problem Formulation

9.2.1 Dynamic System Model

The electrical circuit of the three-phase two-level grid-connected power converter under consideration is shown in Fig. 9.1. The system is connected to the grid through a smoothing inductor L with a parasitic resistance r. It is assumed that an equivalent resistive load R_L is connected at the DC-link capacitor C. The load is considered as an unknown external disturbance.

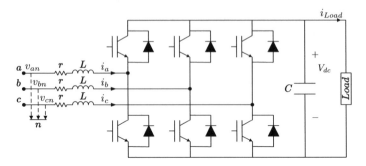

Fig. 9.1 Circuit of the three-phase two-level grid-connected power converter

The grid current and DC-link capacitor voltage dynamics can be written in a dq SRF rotating at the grid frequency [2]. Assuming that grid voltages are balanced then the system model is

$$L\frac{di_d}{dt} = -ri_d + \omega Li_q + v_d - \delta_d V_{dc},\tag{9.1}$$

$$L\frac{di_q}{dt} = -ri_q - \omega Li_d + v_q - \delta_q V_{dc},\tag{9.2}$$

$$C\frac{dV_{dc}}{dt} = \left(\delta_d i_d + \delta_q i_q\right) - i_{Load},\tag{9.3}$$

where δ_d, δ_q are the switching functions, v_d, v_q are grid voltages, i_d, i_q are input currents, V_{dc} is DC-link capacitor voltage and ω is the angular frequency of the grid. From the control point of view, working on the dq SRF has the advantage of reducing the current control task into a set-point tracking problem.

It should be pointed out that the amplitude of the control vector is constrained due to the fact that only implementable control vectors are contained in an area limited by the well-known hexagon. Therefore, it is necessary to consider the constraint $\|\delta_{dq}\| \leq \sqrt{2}$, which implies that the switching function stays unsaturated if its magnitude is not larger than $\sqrt{2}$. This is a conservative approach but ensures the system operation.

9.2.2 Modeling Uncertainties

In practical applications, the system modeling is usually obtained under several assumptions, e.g. ignore the switching losses, in order to simplify the control design. However, such models are not exact, and may vary from real system behavior under some operation conditions. Therefore, a controller should be designed to be robust against any modeling parametric uncertainties. For this reason, parametric uncertainties, i.e. smoothing inductor L, parasitic phase resistance r and the angular frequency of the grid ω, are fully considered and included into the model such that the control robustness is guaranteed. These parameters are formalized as follows:

$$L = L_0 + \Delta L, \; r = r_0 + \Delta r, \; \omega = \omega_0 + \Delta \omega, \tag{9.4}$$

where L_0, r_0, ω_0 are the nominal values and ΔL, Δr, $\Delta \omega$ are the parametric uncertainties which can be considered as unknown slow variant signals.

9.2.3 Control Objectives

The control objectives for three-phase two-level grid-connected power converters are as follows:

- The currents i_d, i_q should track their references i_d^* and i_q^*, respectively. The reference value i_d^* is calculated in such a way that DC-link capacitor voltage is regulated to a certain value. The reference value i_q^* is set to provide a desired instantaneous reactive power.

$$i_d \to i_d^*, \; i_q \to i_q^*. \tag{9.5}$$

- The DC component of the DC-link capacitor voltage should be driven to some reference value V_{dc}^*.

$$V_{dc} \to V_{dc}^*. \tag{9.6}$$

9.3 Disturbance Observer Based SMC Design

In three-phase two-level grid-connected power converters there exist different kinds of disturbances, like parameter uncertainties and load variations. If the controller does not have enough ability to reject these disturbances then they will degrade the performance of closed loop system. A cascade control structure is used to govern the system (9.1)–(9.3). The controller consists of a current tracking loop, inner loop, and an ESO-based voltage regulation loop, outer loop. For the current tracking loop, an STA controller is designed which ensures fast convergence of the currents i_d and i_q to their references i_d^* and i_q^*, respectively. For the voltage regulation loop, an ESO is used to estimate the load power which is considered as an external disturbance. Besides, an STA controller is implemented in parallel to regulate the DC-link capacitor voltage to its desired value using the estimate of the disturbance. In this section, firstly, the basics of STA will be briefly shown. Secondly, the design of both control loops will be presented.

9.3.1   Recalling of Super-Twisting Algorithm

SMC has been developed as a new control design method for a wide spectrum of systems including nonlinear, time-varying and fault tolerant systems [12, 14]. The STA is one of the most popular SOSM algorithms and a unique absolutely continuous sliding mode algorithm, ensuring all the main properties of first order SMC for systems with Lipschitz continuous matched uncertainties/disturbances with bounded gradients [8]. This part discusses the STA in a general case for a single input nonlinear system.

The sliding mode design approach consists of two steps. The first step considers the choice of sliding manifold which provides desired performance in the sliding mode. The second step concerns the design of a control law which will force the system states to reach the sliding manifold in finite time, thus the desired performance is attained and maintained. This part discusses the STA in a general case for a single input nonlinear system.

Consider a nonlinear system

$$
\begin{aligned}
\dot{x} &= a(x) + b(x, u), \\
y &= s(t, x),
\end{aligned}
\tag{9.7}
$$

where $x \in X \subset \mathcal{R}^n$ is the state vector, $u \in U \subset \mathcal{R}$ is the input, $s(t, x) : \mathcal{R}^{n+1} \to \mathcal{R}$ is the sliding variable and $a(x)$ and $b(x, u)$ are smooth uncertain functions.

The control objective is to force s and its time derivative $\dot{s}$ to zero. By differentiating the sliding variable $s(t, x)$ twice, the following relations are derived:

$$
\begin{aligned}
\dot{s} &= \frac{\partial}{\partial t} s(t, x) + \frac{\partial}{\partial x} s(t, x) [a(x) + b(x, u)], \\
\ddot{s} &= \frac{\partial}{\partial t} \dot{s}(t, x, u) + \frac{\partial}{\partial x} \dot{s}(t, x, u) [a(x) + b(x, u)] + \frac{\partial}{\partial u} \dot{s}(t, x, u) \dot{u} \\
&= \varphi(t, x, u) + \gamma(t, x, u) \dot{u}.
\end{aligned}
\tag{9.8}
$$

Assuming that the sliding variable s has relative degree one with respect to the control input u, i.e. $\frac{\partial}{\partial u} \dot{s}(t, x, u) \neq 0$, there exist positive constant values Φ, Γ_m and Γ_M such that the following conditions are satisfied,

$$
0 < \Gamma_m < \gamma(t, x, u) < \Gamma_M,
\tag{9.9}
$$
$$
-\Phi \leq \varphi(t, x, u) \leq \Phi.
\tag{9.10}
$$

Under the conditions (9.9) and (9.10), the following differential inclusion can be obtained:

$$
\ddot{s} \in [-\Phi, +\Phi] + [\Gamma_m, \Gamma_M] \dot{u}.
\tag{9.11}
$$

In the sequel, a control law based on STA is designed. It consists of two terms, one is the integral of its discontinuous time derivative while the other is a continuous function of the available sliding variable s.

$$u = u_1 + u_2, \quad \dot{u}_1 = -\alpha \text{sign}(s), \quad u_2 = -\lambda |s|^{\frac{1}{2}} \text{sign}(s), \tag{9.12}$$

where α and λ are design parameters that can be determined from the boundary conditions (9.9) and (9.10). The sufficient conditions for finite time convergence to the sliding manifold $s = \dot{s} = 0$ are [18]:

$$\alpha > \frac{\Phi}{\Gamma_m}, \quad \lambda^2 \geq \frac{4\Phi}{\Gamma_m^2} \frac{\Gamma_M}{\Gamma_m} \frac{\alpha + \Phi}{\alpha - \Phi}. \tag{9.13}$$

Remark 9.1 In the case of relative degree one systems, traditional first-order sliding mode could also be applied. However, motivated by the chattering elimination aim, super-twisting based control is employed which means that the control signal u is continuous and chattering is avoided.

9.3.2 Extended State Observer Design

According to the control objective (9.6), the outer control loop is designed to regulate the output capacitor voltage to its reference value V_{dc}^*. For the system (9.1)–(9.3), the current dynamics are much faster than the DC-link capacitor voltage dynamics [32]. Under this condition, it can be considered that $i_d \simeq i_d^*$ and $i_q \simeq i_q^*$. Based on the singular perturbation theory [13], if the fast dynamics are stable then (9.3) will be reduced to

$$C \frac{dV_{dc}}{dt} = \frac{1}{V_{dc}} \left(p^* - p_{load} \right), \tag{9.14}$$

where $p^* = v_d i_d^* + v_q i_q^*$ and $p_{load} = V_{dc} i_{load}$.

It should be noted from (9.14) that p_{load} can be considered as an external disturbance. This paper proposes to design an ESO to estimate the disturbance asymptotically. Then, it will be injected into the control design. Unlike traditional observers, such as Luenberger observer [23], high-gain observer [1] and UIO [4], ESO regards the disturbances of the system as new system states which are conceived to estimate not only the external disturbances but also plant dynamics [9, 28, 29].

To design the ESO, the new variable $z = V_{dc}^2/2$ is introduced in (9.14), yielding

$$C\dot{z} = p^* - d(t), \tag{9.15}$$

with $d(t) = p_{load}$. A linear ESO is given by

$$C\dot{\hat{z}} = p^* - \hat{d}(t) + \beta_1(z - \hat{z}), \tag{9.16}$$

$$\dot{\hat{d}}(t) = -\beta_2(z - \hat{z}), \tag{9.17}$$

where the positive gains β_1 and β_2 are chosen such that the polynomial

$$\lambda^2 + \frac{\beta_1}{C}\lambda + \frac{\beta_2}{C}, \tag{9.18}$$

is Hurwitz stable. Therefore, its natural frequency ω_n and damping ratio ξ are:

$$\omega_n = \sqrt{\frac{\beta_2}{C}}, \quad \xi = \frac{\beta_1}{2}\sqrt{\frac{1}{\beta_2 C}}. \tag{9.19}$$

Denote the observation errors $\epsilon_z = z - \hat{z}$, $\epsilon_d = d(t) - \hat{d}(t)$, the error dynamics are given by,

$$C\dot{\epsilon}_z = -\beta_1\epsilon_z - \epsilon_d, \tag{9.20}$$

$$\dot{\epsilon}_d = \beta_2\epsilon_z + h(t), \tag{9.21}$$

where $h(t) = \dot{d}(t)$ is the variation rate of load power. The system (9.20) and (9.21) can be written as follows:

$$\dot{\epsilon} = A\epsilon + \psi, \tag{9.22}$$

where $\epsilon = [\epsilon_z, \ \epsilon_d]^T$, $A = \begin{bmatrix} -\frac{\beta_1}{C} & -\frac{1}{C} \\ \beta_2 & 0 \end{bmatrix}$ and $\psi = [0 \ h(t)]^T$.

Lemma 9.1 *Suppose that h(t) is bounded, there exist a constant $\delta > 0$ and a finite time $T_1 > 0$ such that the trajectories of the system (9.22) are bounded, $\|\epsilon\| \leq \delta$, $\forall t \geq T_1 > 0$.*

Proof Solving (5.15), one has

$$\epsilon(t) = e^{(t-t_0)A}\epsilon(t_0) + \int_{t_0}^{t} e^{(t-\tau)A}\psi(\tau)\,d\tau, \tag{9.23}$$

where t_0 is the initial time. Using the bound $\left\|e^{(t-t_0)A}\right\| \leq ke^{-\beta(t-t_0)}$, with $k > 0$, $\beta > 0$, then (9.23) can be estimated as follows:

$$\|\epsilon(t)\| \leq ke^{-\beta(t-t_0)}\|\epsilon(t_0)\| + \int_{t_0}^{t} ke^{-\beta(t-\tau)}\|\psi(\tau)\|\,d\tau$$

$$\leq ke^{-\beta(t-t_0)}\|\epsilon(t_0)\| + \frac{k}{\beta}\sup_{t_0\leq\tau\leq t}\|\psi(\tau)\|. \tag{9.24}$$

It follows from (9.24) that $\|\epsilon\| \le \delta$, $\forall t \ge T_1 > 0$, where δ is a positive constant that depends on k, β and the upper bound of $\|\psi(\tau)\|$. ∎

Remark 9.1 In view of (9.20) and (9.21), the parameters β_1 and β_2 determine the bandwidth of the ESO, i.e.

$$\omega_b = \omega_n \sqrt{(4\xi^4 - 4\xi^2 + 2)}. \tag{9.25}$$

Generally speaking, the larger the ESO bandwidth is, the more accurate estimation will be achieved. However, this increases the noise sensitivity due to the augment of the bandwidth. The function $h(t)$ represents the rate of change of load power. If this value is quite large then it means that the load power changes very rapidly. In this case, the observer bandwidth needs to be sufficiently large for an accurate estimate of $d(t)$. Therefore, the selection of β_1 and β_2 should balance between the estimation performance and the noise tolerance.

9.3.3 Capacitor Voltage Regulation Loop

Define the regulation error $\tilde{z} = z^* - z$ with $z^* = \left(V_{dc}^*\right)^2/2$, it follows that,

$$C\dot{\tilde{z}} = -p^* + d(t). \tag{9.26}$$

Notice that the perturbation $d(t)$ is an unknown time varying variable. The proposed ESO-based STA controller for the voltage control loop is given by

$$p^* = \mu_{dc}(\tilde{z}) + \hat{d}(t), \tag{9.27}$$

in which $\mu_{dc}(\tilde{z})$ is the STA which takes the following form,

$$\mu_{dc}(\tilde{z}) = \lambda_{dc}|\tilde{z}|^{\frac{1}{2}}\text{sign}(\tilde{z}) + \alpha_{dc}\int_0^t \text{sign}(\tilde{z})d\tau, \tag{9.28}$$

with some positive constants λ_{dc} and α_{dc}. Substituting (9.27) into (9.26), yields

$$C\dot{\tilde{z}} = -\mu_{dc}(\tilde{z}) + \epsilon_d. \tag{9.29}$$

It can be easily obtained from Lemma 9.1 that

$$\|\dot{\epsilon}_d\| \le \|A\| \delta + \sup_{t_0 \le \tau \le t} \|\psi(\tau)\| = F_d, \tag{9.30}$$

with $\|A\| = \sqrt{\lambda_{\max}(A^T A)}$ and F_d is a positive value. The sufficient conditions for the finite time convergence to the sliding manifold $\tilde{z} = \dot{\tilde{z}} = 0$ are [18]:

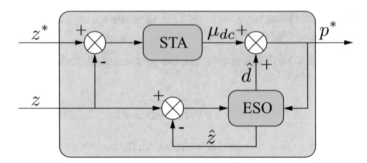

Fig. 9.2 Block diagram of the STA + ESO controller for the voltage loop

$$\alpha_{dc} > CF_d, \quad \lambda_{dc}^2 \geq 4C^2 F_d \frac{\alpha_{dc} + F_d}{\alpha_{dc} - F_d}. \tag{9.31}$$

The structure of the ESO-based voltage regulation loop is shown in Fig. 9.2.

9.3.4 Current Tracking Loop

The objective of the inner control loop is to force i_d and i_q to their references i_d^* and i_q^* respectively. The reference current i_d^* is computed from the output of the outer control loop and is calculated to achieve dynamic voltage regulation. The reference current i_q^* is set to provide a desired instantaneous reactive power q^*. Therefore, the current references are calculated as

$$i_d^* = \frac{p^*}{v_d}, \quad i_q^* = \frac{q^*}{v_d}. \tag{9.32}$$

The sliding mode variables for the current control are defined as,

$$s_d = i_d^* - i_d, \quad s_q = i_q^* - i_q. \tag{9.33}$$

Taking the first time derivative of $s_{dq} = [s_d \ s_q]^\mathrm{T}$ yields,

$$\begin{bmatrix} \dot{s}_d \\ \dot{s}_q \end{bmatrix} = \begin{bmatrix} \dot{i}_d^* + \frac{r}{L}i_d - \frac{v_d}{L} - \omega i_q \\ \dot{i}_q^* + \frac{r}{L}i_q - \frac{v_q}{L} + \omega i_d \end{bmatrix} + \frac{V_{dc}}{L} \begin{bmatrix} \delta_d \\ \delta_q \end{bmatrix}. \tag{9.34}$$

In order to satisfy the saturation constraint, the controllers δ_d and δ_q are designed as follows:

$$\delta_d = \sigma\left(m_d\right), \tag{9.35}$$

$$\delta_q = \sigma\left(m_d\right), \tag{9.36}$$

where

$$m_d = \frac{L_0}{V_{dc}}\left[-\mu_d(s_d) + \frac{v_d}{L_0} - \frac{r_0}{L_0}i_d - \dot{i}_d^* + \omega_0 i_q \right],$$ (9.37)

$$m_q = \frac{L_0}{V_{dc}}\left[-\mu_q(s_q) + \frac{v_q}{L_0} - \frac{r_0}{L_0}i_q - \dot{i}_q^* - \omega_0 i_d \right],$$ (9.38)

$\mu_d(s_d)$ and $\mu_q(s_q)$ are STAs which are in the form

$$\mu_d(s_d) = \lambda_d |s_d|^{\frac{1}{2}} \mathrm{sign}(s_d) + \alpha_d \int_0^t \mathrm{sign}(s_d)d\tau,$$ (9.39)

$$\mu_q(s_q) = \lambda_q |s_q|^{\frac{1}{2}} \mathrm{sign}(s_q) + \alpha_q \int_0^t \mathrm{sign}(s_q)d\tau,$$ (9.40)

with some positive constants λ_i, α_i, $i \in \{d, q\}$, and $\sigma(x)$ is a saturation function.

Theorem 9.1 *Consider the system (9.1)–(9.3) in closed loop with the saturated controller (9.35) and (9.36). This yields*

$$\dot{s}_d = -\frac{L_0}{L}\mu(s_d) + \varphi_d(t),$$ (9.41)

$$\dot{s}_q = -\frac{L_0}{L}\mu(s_q) + \varphi_q(t),$$ (9.42)

where

$$\varphi_d(t) = \frac{\Delta r}{L}i_d + \frac{\Delta L}{L}\dot{i}_d^* + \frac{\omega_0 L_0 - \omega L}{L}i_q,$$ (9.43)

$$\varphi_q(t) = \frac{\Delta r}{L}i_q + \frac{\Delta L}{L}\dot{i}_q^* - \frac{\omega_0 L_0 - \omega L}{L}i_d.$$ (9.44)

The state trajectories of the system (9.41) and (9.42) converge to the origin $s_d = 0$, $s_q = 0$ in finite time if the gains of $\mu(s_d)$, $\mu(s_q)$ and V_{dc}^* are chosen such that the following conditions are satisfied,

$$\alpha_d > \frac{\gamma_d}{1 - \gamma_0}, \quad \lambda_d^2 > \alpha_d,$$
$$\alpha_q > \frac{\gamma_q}{1 - \gamma_0}, \quad \lambda_q^2 > \alpha_q,$$ (9.45)

and

$$V_{dc}^* > \sqrt{2\left(L\omega \|i_{dq}^*\|\right)^2 + 3E^2},$$ (9.46)

where γ_0, γ_d and γ_q are positive constants and E is the amplitude of the grid source.

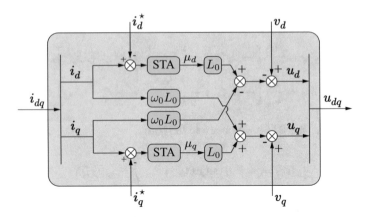

Fig. 9.3 Block diagram of the current tracking regulation loop

Proof The proof is divided into two parts. Firstly, if the control vector $\|\delta_{dq}\| \le \sqrt{2}$, then $\delta_{dq} = [m_d, \ m_q]^{\mathrm{T}}$. Secondly, in the case when the control vector $\|u_{dq}\| > \sqrt{2}$, thus the control vector is saturated to that value.

Case 1: The controller is given by $\delta_d = m_d$ and $\delta_q = m_q$ in (9.35) and (9.36). Considering closed loop behavior (9.41) and (9.42) and given that (9.1)–(9.3) is a physical system, thus it is reasonable to assume that the variables $\varphi_d(t)$ and $\varphi_q(t)$ and its time derivatives are bounded functions:

$$\|\dot{\varphi}_d(t)\| \le \gamma_d, \quad \|\dot{\varphi}_q(t)\| \le \gamma_q, \tag{9.47}$$

with some positive constants γ_d and γ_q. Taking into account that

$$(1 - \gamma_0)L \le L_0 \le (1 + \gamma_0)L, \tag{9.48}$$

for some scalar $0 < \gamma_0 < 1$. Then, according to [18], the trajectories of the system (9.41)–(9.42) will converge to $s_d = \dot{s}_d = 0$ and $s_q = \dot{s}_q = 0$ in finite time when the controller gains satisfy conditions (9.45).

Case 2: According to [6], the sufficient condition for the control vector δ_{dq} to enter into the circle of radium $\sqrt{2}$, (i.e., $\|\delta_{dq}\| \le \sqrt{2}$) is that V_{dc}^* satisfies the condition (9.46). Thus, Theorem 9.1 is proven. ∎

Remark 9.2 It should be noted that the cross-coupling terms and the source voltage are compensated by (9.35) and (9.38). However, (9.37) and (9.38) require the information of the derivative of i_d^* and i_q^*. Usually, these values are zero during the steady state and only affect slightly the system performance during the transient state. Therefore, these terms are neglected in the final control law. Similarly, parameter r_0 is very small in order to reduce system losses. Moreover, this parameter is usually unknown and its use it is also avoided. Taking into account these considerations, the block diagram for the current tracking control loop is shown in Fig. 9.3, where $u_{dq} = V_{dc}\delta_{dq}$.

9.4 Experimental Results

In order to demonstrate the feasibility of the proposed control algorithm, practical results have been performed, comparing the proposed ESO-based SOSM control to a well tuned linear conventional PI-SRF regulator. The PI-SRF is considered as the standard control method for this system and has been selected as a baseline controller. The electrical parameters of the power converter, DC-link voltage reference, and switching and sampling frequencies for the experimental setup are summarized in Table 9.1.

The power converter prototype used for the experiment is shown in Fig. 9.4. A digital implementation of both current and DC-link voltage control algorithms is executed in a TMS320F28335 floating point digital signal processor board. Two sets of experiments are done. The first one consists of a load step at DC-link from no load to full load (3.125 kW). To perform the experiment, the capacitor voltage reference is set to 750 V and a 180 Ω resistive load is suddenly connected to the DC-link. The second test focuses on the reactive power tracking ability. For this purpose, an instantaneous reactive power command step is done. Measurements of DC-link voltage, phase voltages and currents, harmonic contents of currents, active power, reactive power, and power factor have been taken.

The parameters of the PI-SRF and ESO-SOSM controllers are given in Table 9.2. They are chosen so that the current dynamics are much faster than the capacitor output voltage dynamics. It should be noticed that a Phase Locked Loop (PLL) is also implemented in the digital platform in order to work in the SRF [36].

Table 9.1 Electrical system parameters

Parameter	Description
Phase-to-neutral voltage (RMS)	230 V
Grid frequency ω_0	50 Hz
Filter inductor L_0	15 mH
DC-link capacitor C	2800 μF
DC-link voltage reference V_{dc}^*	750 V
Sampling frequency f_s	10 kHz
Switching frequency f_{sw}	10 kHz

Table 9.2 Controller design parameters

PI-SRF	Values
Voltage regulation loop (k_{pdc}, k_{idc})	0.04, 0.5
Current tracking loop (k_{pd}, k_{pq}, k_{id}, k_{iq})	75, 400, 75, 400
ESO-SOSM	Values
Voltage regulation loop (λ_{dc}, α_{dc}, β_1, β_2)	3, 750, 3, 300
Current tracking loop (λ_d, α_d, λ_q, α_q)	85, 20000, 85, 20000

(a)

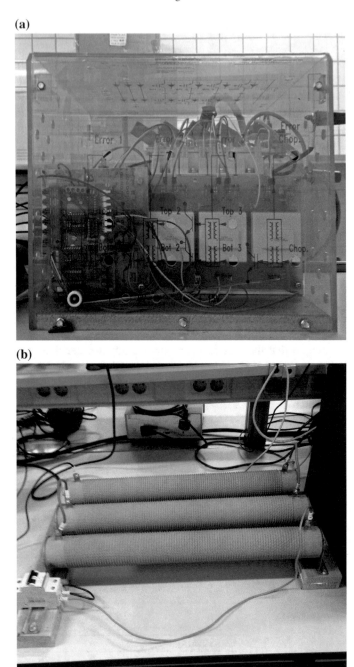

(b)

Fig. 9.4 Power converter prototype: **a** Power converter, **b** DC load

9.4.1 Load Step at DC-Link

The first test consists of evaluating the proposed controller performance under a load step at the DC-link capacitor. Figure 9.5 shows the transient response of the DC-link capacitor voltage for a load step from no load to a load composed by a resistor of 180 Ω. Three waveforms are presented. Figure 9.5a corresponds with the conventional approach with a PI-SRF controller. Figure 9.5b is the result achieved with the proposed ESO-SOSM algorithm and Fig. 9.5c is associated to the PI-SRF but increasing the bandwidth of the controller.

The results show that, both control laws can achieve the DC-link capacitor voltage regulation. Comparing Fig. 9.5a with Fig. 9.5b, the settling time is roughly the same when PI-SRF and ESO-SOSM are used but the last one reduces the DC-link capacitor voltage drop. The PI-SRF results in a voltage drop of 45 V while in the case of the ESO-SOSM approach this is only 22 V. Therefore, the proposed controller reduces the DC-link capacitor voltage drop under a load step in 48.9%. Clearly the ESO-SOSM requires less energy from the DC-link capacitors compared with the PI-SRF. Same DC-link voltage drop can be achieved with the PI-SRF but it is necessary to increase the outer controller bandwidth. It is possible to do this, but it is well known that increasing the bandwidth of the outer control loop in the conventional PI-SRF affects the grid current harmonic content. Table 9.3 collects the current THD computed up to 50th harmonics for the three situations. Clearly the ESO-SOSM allows to achieve a better transient response while maintaining a low THD in steady state.

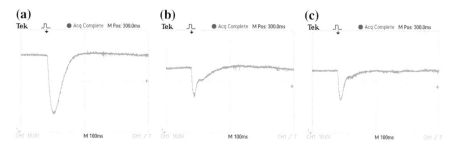

Fig. 9.5 Transient response of DC-link capacitor voltage under a load step: **a** PI-SRF algorithm, **b** ESO-SOSM strategy, **c** PI-SRF algorithm with increased bandwidth

Table 9.3 Grid current THD for the load step experiment

Controller	THD(%)
PI-SRF	2.4
ESO-SOSM	2.1
PI-SRF $((k_{pdc}, k_{idc}) * 3)$	4.9

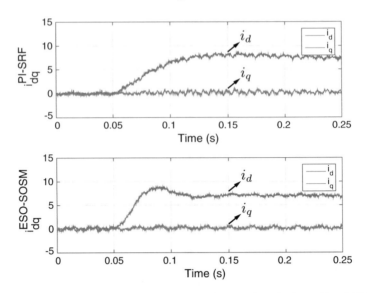

Fig. 9.6 Transient response of grid currents in the dq frame under a load step. Top: PI-SRF algorithm. Bottom: ESO-SOSM strategy

Figure 9.6 plots the grid currents for the load step at the DC-link capacitor tests. The waveforms are represented in the SRF. Only the currents for Fig. 9.5a and b are plotted. Clearly the PI-SRF presents a slow transient response and needs some time to reach the steady state value. On the other hand, the ESO-SOSM approach quickly changes the i_d current component value. It should be noticed that i_d^* comes from the outer control loop. Therefore, the ESO introduced in the control law provides a clear improvement compared with the conventional approach that only considers a PI for the DC-link voltage regulation.

9.4.2   Reactive Power Command Step

The second experiment assesses the control law capability to provide a desired amount of reactive power. To do so, it is introduced a instantaneous reactive power command step from 0 to 3 kVAr. The system response is evaluated for both the PI-SRF and ESO-SOSM controllers.

Figure 9.7 presents the transient state for the experimental results achieved by the PI-SRF and Fig. 9.8 shows the corresponding one to the ESO-SOSM. As expected, both controllers provide the desired reactive power. However, the dynamic response for the ESO-SOSM is faster than the PI-SRF. The performance in most interest is the steady-state behavior. Figure 9.9a–c show the steady state features for the PI-SRF controller. Figure 9.9a–c are the grid currents, the current spectrum and information

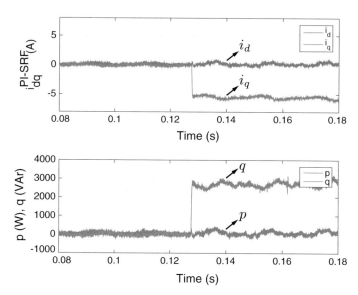

Fig. 9.7 Transient response of grid currents in the dq frame under a load step for the PI-SRF algorithm. Top: Grid currents in the SRF. Bottom: Instantaneous active and reactive power

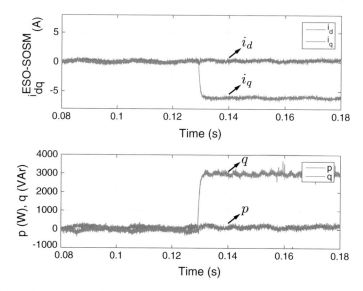

Fig. 9.8 Transient response of grid currents in the dq frame under a load step for the ESO-SOSM strategy. Top: Grid currents in the SRF. Bottom: Instantaneous active and reactive power

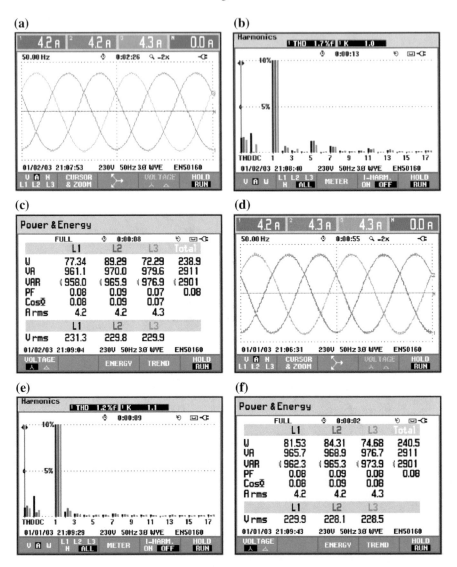

Fig. 9.9 Steady state currents for a instantaneous reactive power command of 3 kVAr: **a** Current waveforms for PI-SRF algorithm, **b** Current spectrum for PI-SRF algorithm, **c** Power factor for PI-SRF algorithm, **d** Current waveforms for ESO-SOSM strategy, **e** Current spectrum for ESO-SOSM algorithm, **f** Power factor for ESO-SOSM strategy

about the active power, reactive power values and the power factor respectively. On the other hand Fig. 9.9e–f present the same information in the case of the ESO-SOSM approach. Comparing the grid spectrum in Fig. 9.9b and e the ESO-SOSM controller has better performance. Particularly, the 5th and 7th harmonics are reduced compared with the PI-SRF results. Furthermore, the current THD is 1.2% for the ESO-SOSM and 1.7% for the PI-SRF. This supposes a reduction of 29.4% when the ESO-SOSM approach is used in the same power converter prototype.

9.5 Conclusions

In this chapter, the problem of disturbance observer based SMC of three-phase two-level grid-connected power converters has been investigated. SOSM technique has been adopted to solve its control problem, which is considered as a promising alternative due to its features of robustness and effectiveness for nonlinear systems.

Considering the dynamics of the output voltage, the load connected to the DC-link capacitor is regarded as a disturbance for the voltage control loop, which directly affects the performance of the whole system. To improve the disturbance rejection ability, a composite control law consisting of SOSM based on STA and disturbance compensation via ESO is proposed for the voltage regulation loop. With the disturbance compensation using ESO, the gains of STA for the outer control loop can be reduced without decreasing the settling time and reducing the DC-link capacitor voltage drop. The current loop has been designed to track the current references in the presence of system parameter uncertainties using an STA. Experimental results have demonstrated that the proposed ESO-SOSM controller performs better than that of the conventional PI SRF control. The power converter operated with the proposed algorithm has achieved less DC-link voltage drop under a sudden load step, shorter settling time and better grid current quality in terms of lower THD and reduced values of low-order harmonics content.

References

1. Atassi, A.N., Khalil, H.K.: A separation principle for the stabilization of a class of nonlinear systems. IEEE Trans. Autom. Control **44**(9), 1672–1687 (1999)
2. Blasko, V., Kaura, V.: A new mathematical model and control of a three-phase AC–DC voltage source converter. IEEE Trans. Power Electron. **12**(1), 116–123 (1997)
3. Carrasco, J.M., Franquelo, L.G., Bialasiewicz, J.T., Galvan, E., Guisado, R.C.P., Prats, M.A.M., Leon, J.I., Moreno-Alfonso, N.: Power-electronic systems for the grid integration of renewable energy sources: a survey. IEEE Trans. Ind. Electron. **53**(4), 1002–1016 (2006)
4. Chen, M.S., Chen, C.C.: Unknown input observer for linear non-minimum phase systems. J. Frankl. Inst. **347**(2), 577–588 (2010)
5. Escobar, G., Chevreau, D., Ortega, R., Mendes, E.: An adaptive passivity-based controller for a unity power factor rectifier. IEEE Trans. Control Syst. Technol. **9**(4), 637–644 (2001)
6. Escobar, G., Ortega, R., Van der Schaft, A.J.: A saturated output feedback controller for the three phase voltage sourced reversible boost type rectifier. In: Proceedings of the 24th Annual Conference of the IEEE Industrial Electronics Society (IECON), vol. 2, pp. 685–690 (1998)
7. Gao, W., Hung, J.C.: Variable structure control of nonlinear systems: a new approach. IEEE Trans. Ind. Electron. **40**(1), 45–55 (1993)

8. Gonzalez, T., Moreno, J.A., Fridman, L.: Variable gain super-twisting sliding mode control. IEEE Trans. Autom. Control **57**(8), 2100–2105 (2012)
9. Han, J.: From PID to active disturbance rejection control. IEEE Trans. Ind. Electron. **56**(3), 900–906 (2009)
10. Harnefors, L., Yepes, A.G., Vidal, A., Doval-Gandoy, J.: Passivity-based controller design of grid-connected VSCs for prevention of electrical resonance instability. IEEE Trans. Ind. Electron. **62**(2), 702–710 (2015)
11. Houari, A., Renaudineau, H., Martin, J.P., Pierfederici, S., Meibody-Tabar, F.: Flatness-based control of three-phase inverter with output LC filter. IEEE Trans. Ind. Electron. **59**(7), 2890–2897 (2012)
12. Hung, J.Y., Gao, W., Hung, J.C.: Variable structure control: a survey. IEEE Trans. Ind. Electron. **40**(1), 2–22 (1993)
13. Khalil, H.K.: Nonlinear Systems, 3rd edn. Prentice Hall (2001)
14. Koren, I., Krishna, C.M.: Fault-Tolerant Systems. Elsevier (2010)
15. Lee, D.C., Lee, G.M., Lee, K.D.: DC-bus voltage control of three-phase AC/DC PWM converters using feedback linearization. IEEE Trans. Ind. Appl. **36**(3), 826–833 (2000)
16. Lee, T.S.: Input-output linearization and zero-dynamics control of three-phase AC/DC voltage-source converters. IEEE Trans. Power Electron. **18**(1), 11–22 (2003)
17. Leon, J.I., Vazquez, S., Franquelo, L.G.: Multilevel converters: control and modulation techniques for their operation and industrial applications. Proc. IEEE **105**(11), 2066–2081 (2017)
18. Levant, A.: Sliding order and sliding accuracy in sliding mode control. Int. J. Control **58**(6), 1247–1263 (1993)
19. Levant, A.: Higher-order sliding modes, differentiation and output-feedback control. Int. J. Control **76**(9–10), 924–941 (2003)
20. Linares-Flores, J., Méndez, A.H., García-Rodríguez, C., Sira-Ramírez, H.: Robust nonlinear adaptive control of a boost converter via algebraic parameter identification. IEEE Trans. Ind. Electron. **61**(8), 4105–4114 (2014)
21. Liu, J., Laghrouche, S., Wack, M.: Observer-based higher order sliding mode control of power factor in three-phase AC/DC converter for hybrid electric vehicle applications. Int. J. Control **87**(6), 1117–1130 (2014)
22. Liu, J., Vazquez, S., Gao, H., Franquelo, L.G.: Robust control for three-phase grid connected power converters via second order sliding mode. In: IEEE International Conference on Industrial Technology (ICIT), pp. 1149–1154 (2015)
23. Luenberger, D.G.: Observing the state of a linear system. IEEE Trans. Mil. Electron. **8**(2), 74–80 (1964)
24. Pahlevaninezhad, M., Das, P., Drobnik, J., Jain, P.K., Bakhshai, A.: A new control approach based on the differential flatness theory for an AC/DC converter used in electric vehicles. IEEE Trans. Power Electron. **27**(4), 2085–2103 (2012)
25. Pan, C.T., Chen, T.C.: Modelling and analysis of a three phase PWM AC–DC convertor without current sensor. IEE Proc. B (Electric Power Appl.) **140**(3), 201–208 (1993)
26. Romero-Cadaval, E., Spagnuolo, G., Franquelo, L.G., Ramos-Paja, C.A., Suntio, T., Xiao, W.M.: Grid-connected photovoltaic generation plants: components and operation. IEEE Ind. Electron. Mag. **7**(3), 6–20 (2013)
27. Shtessel, Y., Baev, S., Biglari, H.: Unity power factor control in three-phase AC/DC boost converter using sliding modes. IEEE Trans. Ind. Electron. **55**(11), 3874–3882 (2008)
28. Sun, L., Li, D., Lee, K.Y.: Enhanced decentralized PI control for fluidized bed combustor via advanced disturbance observer. Control Eng. Pract. **42**, 128–139 (2015)
29. Sun, L., Li, D., Zhong, Q.C., Lee, K.Y.: Control of a class of industrial processes with time delay based on a modified uncertainty and disturbance estimator. IEEE Trans. Ind. Electron. **63**(11), 7018–7028 (2016)
30. Tan, S.C., Lai, Y.M., Tse, C.K., Martinez-Salamero, L., Wu, C.K.: A fast-response sliding-mode controller for boost-type converters with a wide range of operating conditions. IEEE Trans. Ind. Electron. **54**(6), 3276–3286 (2007)

31. Thounthong, P.: Control of a three-level boost converter based on a differential flatness approach for fuel cell vehicle applications. IEEE Trans. Veh. Technol. **61**(3), 1467–1472 (2012)
32. Umbria, F., Aracil, J., Gordillo, F., Salas, F., Sanchez, J.A.: Three-time-scale singular perturbation stability analysis of three-phase power converters. Asian J. Control **16**(5), 1361–1372 (2014)
33. Vazquez, S., Leon, J.I., Franquelo, L.G., Rodriguez, J., Young, H.A., Marquez, A., Zanchetta, P.: Model predictive control: a review of its applications in power electronics. IEEE Ind. Electron. Mag. **8**(1), 16–31 (2014)
34. Vazquez, S., Liu, J., Gao, H., Franquelo, L.G.: Second order sliding mode control for three-level NPC converters via extended state observer. In: 41st Annual Conference of the IEEE Industrial Electronics Society (IECON), pp. 005118–005123 (2015)
35. Vazquez, S., Marquez, A., Aguilera, R., Quevedo, D., Leon, J.I., Franquelo, L.G.: Predictive optimal switching sequence direct power control for grid-connected power converters. IEEE Trans. Ind. Electron. **62**(4), 2010–2020 (2015)
36. Vazquez, S., Sanchez, J.A., Reyes, M.R., Leon, J.I., Carrasco, J.M.: Adaptive vectorial filter for grid synchronization of power converters under unbalanced and/or distorted grid conditions. IEEE Trans. Ind. Electron. **61**(3), 1355–1367 (2014)
37. Wang, G., Konstantinou, G., Townsend, C.D., Pou, J., Vazquez, S., Demetriades, G.D., Agelidis, V.G.: A review of power electronics for grid connection of utility-scale battery energy storage systems. IEEE Trans. Sustain. Energy **7**(4), 1778–1790 (2016)
38. Wu, R., Dewan, S.B., Slemon, G.R.: A PWM AC-to-DC converter with fixed switching frequency. IEEE Trans. Ind. Appl. **26**(5), 880–885 (1990)

Printed in the United States
By Bookmasters